T0329610

Responding to Climate Change

KDI/EWC SERIES ON ECONOMIC POLICY

The Korea Development Institute (KDI) was established in March 1971 and it is Korea's oldest and best-known research institute in the fields of economic and social sciences. Its mission is to contribute to Korea's economic prosperity by drafting socioeconomic development plans and providing timely policy recommendations based on rigorous analysis. Over the decades KDI has effectively responded to rapidly changing economic conditions at home and abroad by conducting forward-looking research as well as putting forth significant efforts in formulating long-term national visions.

The East-West Center promotes better relations and understanding among the people and nations of the United States, Asia and the Pacific through cooperative study, research, and dialogue. It serves as a resource for information and analysis on critical issues of common concern, bringing people together to exchange views, build expertise and develop policy options. The Center is an independent, public, nonprofit organization with funding from the U.S. government, and additional support provided by private agencies, individuals, foundations, corporations, and governments in the region.

The KDI/EWC series on Economic Policy aims to provide a forum for scholarly discussion, research and policy recommendations on all areas and aspects of contemporary economics and economics policy. Each constituent volume in this series will prove invaluable reading to a wide audience of academics, policymakers and interested parties such as NGOs and consultants.

Responding to Climate Change

Global Experiences and the Korean Perspective

Edited by

Chin Hee Hahn

Kyungwon University, Korea

Sang-Hyop Lee

East-West Center and University of Hawaii at Manoa, USA

Kyoung-Soo Yoon

Korea Development Institute, Seoul, Korea

KDI/EWC SERIES ON ECONOMIC POLICY

A JOINT PUBLICATION OF THE KOREA DEVELOPMENT
INSTITUTE, THE EAST-WEST CENTER, AND EDWARD ELGAR
PUBLISHING LTD

Edward Elgar

Cheltenham, UK • Northampton, MA, USA

Published by
Edward Elgar Publishing Limited
The Lypiatts
15 Lansdown Road
Cheltenham
Glos GL50 2JA
UK

Edward Elgar Publishing, Inc.
William Pratt House
9 Dewey Court
Northampton
Massachusetts 01060
USA

A catalogue record for this book
is available from the British Library

Library of Congress Control Number: 2011936419

MIX
Paper from
responsible sources
FSC
www.fsc.org   FSC® C018575

ISBN 978 0 85793 995 1

Typeset by Servis Filmsetting Ltd, Stockport, Cheshire
Printed and bound by MPG Books Group, UK

Contents

Contributors

Richard A. Bradley is the Head of the Energy Efficiency and Environment Division at the International Energy Agency.

Chin Hee Hahn is a former Senior Fellow of the Korea Development Institute and an Associate Professor of Economics at Kyungwon University.

Stephen Howes is a Professorial Fellow of the Crawford School of Economics and Government at the Australian National University.

Sang-Hyop Lee is a Fellow at the East-West Center and an Associate Professor of the Department of Economics at the University of Hawaii at Manoa.

Jin-Gyu Oh is the Managing Director of the Green Growth Research Group at the Korea Energy Economics Institute.

Lawrence Rothenberg is the Corrigan-Minehan Professor of Political Science of the Department of Political Science at the University of Rochester.

Sung-Hyun Ryu is a Researcher at the Korea Development Institute.

J.P.M. (Jos) Sijm is a Researcher of the Department of Policy Studies at the Energy Research Centre (ECN) of the Netherlands.

Min-Kyu Song is a Research Fellow of the Financial Markets and System Division at the Korea Institute of Finance.

Kyoung-Soo Yoon is a Research Fellow of the Department of Industrial and Corporate Affairs at the Korea Development Institute.

ZhongXiang Zhang is a Senior Fellow at the East-West Center and an Adjunct Professor at the Chinese Academy of Sciences and Peking University.

Abbreviations and acronyms

Annex I	the 37 developed countries and countries in transition committed under the 1997 Kyoto protocol to the UNFCCC to reducing greenhouse gases
BAU	business as usual
CAFE	corporate average fuel economy (a U.S. government standard)
CCP	centralized counter parties
CDM	clean development mechanism
CITL	Community Independent Transaction Log
CO_2	carbon dioxide
CO_2eq	carbon dioxide equivalent
DSM	demand side management
EAR	emissions allowance requirements
EEA	Energy Exchange Austria
EEX	European Energy Exchange
EPA	United States Environmental Protection Agency
ETS	emissions trading scheme
E.U.	European Union
EU-27	European Union (all 27 member countries)
EU ETS	European Union Emissions Trading Scheme
EUA	E.U. allowance
EXX	European Climate Exchange
GATT	General Agreement on Tariffs and Trade
GDP	gross domestic product
GHG	greenhouse gas
GTEM	Global Trade and Environment Model
GW	gigawatt
GWh	gigawatt-hour
IEA	International Energy Agency
IPCC	Intergovernmental Panel on Climate Change
IT	information technology
JI	joint implementation
LNG	liquefied natural gas
MAC	marginal abatement cost

$mtCO_2$	million tons of carbon dioxide
$mtCO_2eq$	million tons of carbon dioxide equivalent
MW	megawatt
MWh	megawatt-hour
NAP	national allocation plan
NATO	North Atlantic Treaty Organization
NGO	nongovernmental organization
OECD	Organisation for Economic Co-operation and Development
OTC	over the counter
PPM	parts per million
R&D	research and development
RD&D	research, development, and demonstration
tCO_2	ton (metric) of carbon dioxide
tCO_2eq	ton (metric) of carbon dioxide equivalent
toe	ton (metric) of oil equivalent
TW	terawatt
TWh	terawatt-hour
U.K.	United Kingdom
U.N.	United Nations
UNFCCC	United Nations Framework Convention on Climate Change
U.S.	United States
US$	United States dollars
VA	voluntary agreement (Korean government program)
WDI	*World Development Indicators* (World Bank)
WTO	World Trade Organization

Preface

Climate change is one of the most important issues in the world economy nowadays. Without significant reduction of greenhouse gas emissions, it will hinder economic development and the sustained growth of human-kind. Confronting this global problem, governments (mostly of developed countries) have been making efforts to mitigate and adapt to it. These efforts have launched international collaboration, resulting in the United Nations Framework Convention on Climate Change (UNFCCC) and the 1997 Kyoto protocol. There are ongoing negotiations for the post-Kyoto era beyond 2012, when more countries are expected to participate in the collective action. Accordingly, many countries are introducing, expanding, or improving their policy measures to respond to climate change.

It appears, however, that many obstacles lie ahead in achieving an efficient and equitable mitigation policy framework, on both the global and the national levels. Owing to different economic conditions and the "public good" nature of climate change, procedures to secure global agreement are being considerably delayed. Domestic policies are often distorted because of conflicts of interest between different industries and regions, and between firms and taxpayers. Political procedures for environmental regulation raise problems of dynamic consistency and commitment.

These challenges are more severe in developing and newly developed countries such as Korea. Concerns that regulation of greenhouse gas emissions could seriously halt economic growth make the adoption of mitigation policy difficult. Governments are faced with difficult considerations when trying to take a balanced position between economic development and environmental protection. Traditional industrial policies and regulations often conflict with emissions regulation and support for "green industries" and "green technologies." Most of all, insufficient experience with environmental issues hampers appropriate policy designs and public consultations.

Under these circumstances, the Korean government declared its Green Growth National Strategy in 2008. The strategy aims to harmonize sustained economic growth and the enhancement of the quality of life, at the same time participating in global cooperation in responding to climate change. Challenges still remain, and the success of this ambitious strategy depends on preparing a policy framework composed of balanced, efficient, and acceptable measures.

This book addresses major climate change issues and reviews policy options. It collects lessons from global experience and discusses the mitigation policy framework of Korea in its initial stage. Particular focus is on the need for a mitigation policy, the design of an emissions trading scheme (one that is known to be an efficient market-based mitigation policy), energy policy in the context of the mitigation framework, the political economy of climate change, and potential conflicts between the international trade framework and the mitigation policy measures. Authors from the United States, the Netherlands, and Australia address and discuss these issues based on global experience of responding to climate change. Authors from Korea review climate change issues from the perspective of Korea, assess policy measures adopted or proposed by the government, and suggest policy directions.

The chapters of this volume are based on papers presented at the conference on Climate Change and Green Growth: Korea's National Growth Strategy, organized by the East-West Center (EWC) and the Korea Development Institute (KDI) in Honolulu, Hawaii, in July 2008. At the EWC-KDI conference, renowned scholars presented and discussed various issues on climate change. The chapters have been developed through fruitful discussions during the conference and updated further by the authors.

On publishing this volume, I would like to thank Dr. Chin Hee Hahn, Associate Professor of Kyungwon University, Dr. Sang-Hyop Lee, Fellow of the EWC and Associate Professor of the University of Hawaii at Manoa, and Dr. Kyoung-Soo Yoon, Research Fellow of the KDI, for coordinating the conference and preparing this volume for publication. My gratitude also goes to Kennon Breazeale at the EWC and Nanhee Kim at the KDI for production coordination. Finally but not least, I also wish to thank the authors of the papers, the discussants, and other participants who contributed to the conference.

Oh-Seok Hyun
President
Korea Development Institute

1. Responding to climate change: introduction and overview

Chin Hee Hahn, Sang-Hyop Lee, and Kyoung-Soo Yoon

Climate change is becoming widely recognized as one of the most important changes in the world economic environment. Taking action to mitigate it has already become an inevitable task for policymakers. According to a forecast of the Organisation for Economic Co-operation and Development (OECD 2009), under the business-as-usual scenario and in the absence of new policy action, the atmospheric concentrations of the world's greenhouse gases (GHGs) will increase to about 650 parts per million (ppm) in 2050, which could cause the global average temperature to be at least 2° C higher than it was in preindustrial times. This in turn would impose a fundamental constraint on economic activities of mankind. Stern (2007) forecasts that the per capita loss from climate change could be about 14.4 percent of the world's average per capita income. Based on this recognition, the World Bank (2008) notes that climate change is one of the "new global trends" that could threaten the sustained growth of the world economy and in particular the growth of developing countries.

Climate change is regarded as a global trend, not only because of the threat it poses, but also because of the wide scope of its influence on economic activities. To mitigate and adapt to climate change, most countries will have to adopt new policy instruments, which may include a carbon tax, a cap-and-trade system, schemes to expand supplies of renewable energy, and tools for improving energy efficiency. In the course of implementing such policies, relative prices of goods and services will change, and markets related to climate change—the so-called green markets—will expand. Consumers will change their consumption patterns, and firms will have to adjust themselves to the new regulatory framework and new market situations. In the sense that economies should persistently improve, or reform, their systems of coping with internal and external environmental changes, as noted by Rodrik (2005), governments will have to take this trend into consideration in policies ranging widely from industry and the regulatory system to development strategy.

There are widespread uncertainties in the trend, however, and one critical source of uncertainty is the outcome of international cooperation and negotiations. Owing to the accumulative, nonassimilative nature of GHG, as well as its characteristic as a public good, international cooperation is essential in response to climate change. Since the 1990s, efforts to counter global warming have gradually intensified worldwide, leading to the adoption of the United Nations Framework Convention on Climate Change (UNFCCC) in 1992 and its enforcement in 1994, followed by the adoption of the Kyoto protocol in 1997 and its enforcement in 2005 for the period up to 2012. Ongoing negotiations to build a post-2012 global climate regime include the Bali Action Plan in 2007 and the Cancun Agreement in 2010, among others. They have, however, left a number of issues unresolved, especially between developed and developing countries. Although some issues were settled in the Cancun Agreement, which incorporates major decisions taken at the time of the 2009 Copenhagen Accord, critical issues such as a mitigation target remain to be solved.

Another critical uncertainty is the development of "green technology," which will affect mitigation costs. The costs, in turn, will affect decisions about whether to adopt a mitigation policy and what the policy design should be. Firms may delay, or even give up, investments in essential technologies if success and profitability seem doubtful. Public investment might be a solution, but any serious concern about policy failure might then create obstacles. This problem will probably be more critical in developing countries, in cases where technology is still at a lower level, although these countries could compete with developed ones in expanding green markets.

Facing the need for mitigation and the uncertainties surrounding it, the Korean government announced its Green Growth National Strategy in 2008, to be implemented according to the newly promulgated Framework Act on Low Carbon Green Growth. The purpose of this law is to harmonize the continued development of the national economy and enhancement of the quality of life, while at the same time fulfilling the nation's international responsibilities for protecting the environment. In other words, the strategy aims at forming a virtuous circular structure of economic development, improvement of the environment, and participation in international cooperation.

It seems appropriate for the Korean government to take the task of the response to climate change more seriously, leaving out the issue of whether green growth can be a national strategy. Foremost of all, the Korean peninsula itself is already experiencing the impact of climate change.

Table 1.1   CO$_2$ emissions statistics for selected regions and countries,
          2007

Region or country	CO$_2$ emissions (million tons)	Rate of increase (%, annual average 1990–2007)	GDP (US$ billion at 2000 PPP)	Emissions per unit of GDP (tons/US$ billion)	Emissions per capita (tons)
World	28962.4	1.91	61428.0	0.47	4.38
OECD	13000.8	0.95	32360.9	0.40	10.97
Annex I	14259.1	0.15	32627.2	0.44	11.21
United States	5769.3	1.01	11468.0	0.50	19.10
Japan	1236.3	0.88	3620.2	0.34	9.68
Canada	572.9	1.67	1046.9	0.55	17.37
China	6071.2	6.03	10155.8	0.60	4.58
Korea	488.7	4.55	1065.7	0.46	10.09

Source: IEA (2009).

According to the National Institute of Meteorological Research (NIMR) of Korea, the average temperature of the six biggest cities increased by 1.7°C between 1912 and 2008, which is higher (even after taking urbanization effects into account) than the global average of 0.71–0.77° C. NIMR also reported that, under the scenario of 720 ppm in 2050, the average temperature at the end of the twenty-first century will be 4° C higher than it was at the end of the twentieth century, and that the volume and volatility of precipitation would increase critically (NIMR 2009).

The need for participation in the international mitigation framework is also increasing. Under the UNFCCC, Korea is not considered a developed country, is not in Annex I (the original list of 37 industrialized countries and economies in transition), and therefore is not responsible for fulfilling the GHG emission reduction target to which Annex I countries are committed. However, taking the size of the Korean economy and its GHG emissions into account (see Table 1.1), it is inevitable for Korea to take some mitigation actions, although such actions could be different from those taken by the more developed countries.

In addition, an increasingly growing market for related industries highlights the need for active measures. Although there could be a debate on whether green technologies and green markets actually stimulate economic growth, the Korean government at least, facing a decline in the potential growth of the economy, seems to view them as a "new engine of growth."

Various questions about coping with climate change, at both the global and the individual country levels, still remain. The most critical question is whether it is worthwhile to pursue climate change mitigation, if various uncertainties are considered. The answer to this question may be simple at the global level, but at the country level it could be quite complicated, owing to strategic considerations and political problems. If the answer is positive, more or less, the next question might be what is the best policy, or the best set of policies, to reduce GHG. Even though people may think that they know the best policy, how to implement it is a different issue, given political and other constraints. Thus, the question of constraints against forming an efficient mitigation framework would follow.

This book is arranged to answer these questions, both from the global perspective and from the standpoint of Korea. The authors do not pursue comprehensive answers for them, nor do they cover all relevant issues. Rather, the book attempts to address some critical issues, and each author presents a careful analysis of one such issue. These analyses provide at least some pieces of the puzzle as a contribution toward a more comprehensive picture.

The first half of the book examines issues from general and global perspectives, focusing on the costs and benefits of climate change mitigation (Chapter 2), the experience of mitigation policy in Europe (Chapter 3), energy policy for mitigation (Chapter 4), political economy in response to climate change (Chapter 5), and potential conflict between mitigation policy and the free trade framework (Chapter 6). The remaining chapters examine problems from the viewpoint of Korea, covering energy dependency and terms of trade (Chapter 7), energy policy (Chapter 8), and the GHG reduction policy of Korea (Chapter 9). We present an overview of each chapter here.

BENEFITS AND COSTS OF MITIGATION

Chapter 2 by Stephen Howes is motivated by the argument that the mitigation cost could be much larger than expected and thus could overwhelm the benefits by holding back economic growth. The main question that the author raises is whether the estimates of the mitigation costs in standard modeling exercises are underestimated, but his analysis provides broader outlines for the benefits and the costs of mitigation.

The estimates of global mitigation costs surveyed by the chapter range from 1 percent to 4 percent of global output in 2050, depending mainly on the modeling method and the stabilization target of the atmospheric concentration of GHG. The costs in this range are modest, corresponding to 0.02–0.10 percentage point of annual average growth decline, or 4 to 17 months of growth delay. The author argues that these estimates are certainly sufficient to justify even a stringent mitigation target. His next question is therefore whether these estimates are reliable.

Howes examines several factors that potentially cause underestimation of the mitigation costs. He first notes that the cost of mitigation might be overestimated, as it stimulates growth through the first-mover advantage or investment and innovation. However, the gain is very uncertain, and industrial policies to pursue the first-mover advantage could merely add further costs. Second, transitional and distributional costs of mitigation would increase the political costs. This may raise the need for structural adjustment programs to accompany mitigation policy. But it does not lead to an underestimation of costs nor to a reduction in the need for mitigation. Third, incomplete participation in the global mitigation framework may increase the overall cost, owing to the limitations of the emissions trading systems, and will reduce the global environmental benefit as a result of "carbon leakage." The effect on the domestic cost of mitigation is ambiguous, however, since it depends on feedback from global mitigation to the domestic economy. The actual problem arising from incomplete participation is that it discourages mitigation efforts by participating countries.

The most critical causes of the underestimation of the costs, the author argues, are poor policies and differences between the costs at global and national levels. Recent experience shows that policymakers may choose inefficient policy options. The additional costs from policy errors should be a cause for alarm, even after taking into consideration the first-mover advantage and multiple market failures inherent in climate change mitigation. The author also emphasizes that decisions to mitigate are made at the national level. In the globally efficient mitigation framework, the mitigation costs vary across countries, owing mainly to the differences in current emission intensities. The cost is mostly higher in developing countries, but it is unlikely that those countries can achieve GHG reductions without appropriate compensatory transfers.

Howes concludes that the national costs of mitigation may well lie above the estimated range in standard modeling exercises. Still, he argues that these factors do not reduce the need for the mitigation, while they could discourage the efforts for it. Rather, what is needed is to reduce the cost and to enhance the environmental return, by designing efficient policies and by assisting developing countries with transfers.

THE DESIGN OF MITIGATION POLICY

What does an efficient mitigation policy look like? Under the Kyoto protocol, a flexible mechanism composed of an emissions trading scheme (ETS), joint implementation (JI), and a clean development mechanism (CDM) has been proposed as a cost-effective means of mitigation. In the mechanism, ETS is the main policy instrument, in that it is directly connected to the emissions reduction target. Up to now, only the European Union has operated an ETS, and its experience provides valuable lessons for the design of efficient mitigation policies by other countries. Chapter 3 by Jos Sijm reviews and evaluates the pilot phase (2005–07) of the EU ETS.

Sijm summarizes major features of the EU ETS from the pilot phase to the second trading period (2008–12). He discusses its cap-and-trade system, unrestricted intraperiod banking and borrowing, the interperiod ban on banking, limited participation in specific sectors, free allocation of emissions allowances, linking JI and CDM credits to the EU ETS, and the operation of the EU ETS registry system. He also surveys a number of studies about the effect of the scheme on emissions reduction, economic growth, industrial competitiveness, the development of a trading market, excessive allocation, and carbon leakage.

The pilot period of the EU ETS revealed several problems resulting from its allocation system, contrary to the theoretical prediction that the cap-and-trade system is capable of achieving cost efficiency, regardless of the allowance allocation method. The decentralized structure of the decisionmaking with almost all free allocation, in addition to relatively short allocation and trading periods, led to inefficient allocation of the emissions reduction target between the ETS and the non-ETS sectors, a race to the bottom, distortion of equity and competition, windfall profits, distortion of incentives for investments in emissions facilities, rent-seeking behavior, and growing uncertainty. Even so, the author argues that, despite a certain degree of excessive allocation during this period, GHG indeed appears to have been reduced. In addition, contrary to arguments that the system could hinder economic growth, degrade the competitiveness of several industries, and trigger carbon leakage, there has been no evidence to support these concerns. Furthermore, despite high price volatility in this period, the emissions trading market grew and expanded remarkably, succeeded in the establishment of the market infrastructure, and encouraged the development of the JI and CDM market.

Based on the lessons and experience from the first phase, the EU ETS is expected to seek institutional improvements after 2012, such as stricter and unified allowance allocation, integrated regulations on allocation, expansion of auction allocation, and extension of coverage and trading periods. Sijm concludes that, despite its defects, the initial phase of the EU ETS can be assessed as successful, in that it laid a foundation for subsequent institutional improvements. More importantly, it established a cultural change that made this new institution more acceptable, and it internalized the carbon price into economic decisionmaking.

The energy sector, owing mainly to its large share of GHG emissions, is of key importance in fulfilling the reduction target. In addition, it has policy goals other than mitigation (energy security being a notable one) and has a distinctive industrial and capital structure. For these reasons, special attention is given to the energy sector in designing a mitigation policy. Do we need a separate energy policy for mitigation in addition to the market based emissions policy such as the ETS? If so, what kinds of policy measures are needed? Richard Bradley addresses this issue in Chapter 4.

Chapter 4 begins with the examination of the proposed emissions reduction target. The goal of stabilizing the GHG concentration at 450 ppm implies that net GHG emissions should peak and start to decline before 2020. This requires a radical change in the current energy infrastructure, including everything related to energy consumption. Considering, however, the lifetime of the capital composing the current energy infrastructure, it may not be easy for investors to expect a rapid transformation of the current capital structure.

Bradley argues that, recognizing such difficulties, it is more important to map out practical policies that contain an incentive system to drive long-term change, while setting the minimum cost as a target at the present. He suggests three elements to consider for planning such practical policies. First, it is necessary to focus on an energy efficiency policy, considering its cost merit and its immediate impact on final energy consumption. Second, the focus of the international discussion as to emissions control needs to be on major emitters and emissions sources, which is transitory in nature. For the present, confining the scope of the emissions control group to major emitters simplifies pending problems, thereby contributing to the process of reaching a global agreement, and also works effectively in terms of environmental protection. Furthermore, considering the huge share of electricity generation in the GHG emissions, it may also be possible to confine the emissions control sector to electricity generation. Lastly, without the development of new technology, it will be impossible to reach the stabilization target of 450 ppm. This highlights the need for efforts to increase research, development, and demonstration (RD&D) in the public sector, which has been stagnant since the early 1990s. In this regard, intergovernment cooperation at the early R&D stage (before the market competition) should be fully recognized, since it could help save resources and alleviate problems of intellectual property rights that could hinder the worldwide proliferation of a new technology.

CONSTRAINTS AGAINST EFFICIENT MITIGATION POLICY

Many factors are involved in climate change and its mitigation, and some of them could actually affect the design of mitigation policies. A policy that is efficient in the mitigation policy framework could conflict with other rules in the economy. More importantly, conflicts of interest among stakeholders and resulting politics could be hurdles to the design of efficient and equitable policies, as already noted in Chapter 2. Two chapters discuss some of the factors constraining policy design.

In Chapter 5, Lawrence Rothenberg reviews the political economy of climate change at both the international and the national levels. As is well known, the principle of policy solutions for climate change is to make the producers of GHG pay for the costs of their actions. Besides this straightforward principle, there have been so far a number of efforts searching for global collective action and advances in green technologies, providing reasons to be optimistic about our ability to deal with this problem effectively. Rothenberg emphasizes, however, that formulating an effective global solution and reflecting it back to domestic policies for implementation is not an easy task in reality. As a result, the atmospheric concentration of GHG continues to rise at an alarming rate.

Rothenberg addresses some key factors related to the political economy of climate change. First, the nature of "climate" or "climate change mitigation" as public goods (in particular, nonexclusion) creates conditions that make it difficult to produce incentives for participation, either by a state or by an individual economic player, in sharing the burden instead of pursuing a "free ride" strategy. Second, advanced economies and developing countries are placed in different positions regarding potential damages from climate change and the domestic demand for a solution to them. To draw developing countries toward international cooperation for climate change mitigation requires wealth transfers from advanced economies, of the sort that China has been advocating. But it is very difficult to expect political leaders of advanced economies to adopt and carry out such policies in their own countries. Their political life depends more on economic growth than on environment. Meanwhile, implementing policies for climate change mitigation, as a way to avoid such a situation, would often cause problems of time consistency and credibility.

Rothenberg suggests that, in order to overcome these political-economic problems, it is necessary for the citizens to be convinced that climate change is real and that climate change mitigation is of considerable personal importance. In addition, political leaders should reflect any international agreement, such as a post-Kyoto protocol, in their domestic policies in a time-consistent manner. Finally, it is essential for the international community to establish an efficient cooperation and monitoring system.

ZhongXiang Zhang in Chapter 6 considers the relationship between mitigation policy and free trade rules, including the potential conflicts and synergies between them. In the current international cooperation framework, the regulation of GHG emissions varies across countries. This disparity is expected to continue for the time being, and it could result in carbon leakage owing to differences in production costs. Furthermore, political pressure would likely be exerted by industries subject to the regulations, because of worries about weakened industrial competitiveness.

Several measures to deal with the disparity have been proposed, and one of them is a border adjustment. In the United States, border adjustment measures in the form of emissions allowance requirements (EAR) are included under the proposed cap-and-trade regime. The author argues that these are the most concrete, unilateral trade measures for leveling the carbon playing field. Under EAR, importers are required to acquire and surrender emissions allowances for goods from countries that have not taken climate actions. The question raised is whether such measures could disturb the world trade order and trigger a trade war.

Zhang focuses on the legality of unilateral EAR under World Trade Organization (WTO) rules. He views border adjustment measures as essential, if regulations on GHG emissions are to be adopted in U.S. legislation. He argues, however, that a conflict between the trade and mitigation regimes could harm both regimes. Therefore, in designing such measures, the United States needs to scrutinize potential conflicts with WTO rules carefully, to make the U.S. legislation comply with the rules. He also suggests that the United States needs to explore cooperative sectoral approaches at an international level.

In addition, the author provides suggestions both for the United States and for its developing-country trade partners. On the U.S. side, for a successful WTO defense of the border adjustment provision, there should be a period of good faith efforts, to reach agreements with the countries concerned, before imposing such trade measures. Second, alternative options, with similar functions but less inconsistent with WTO provisions, should be considered. Finally, submissions by importers of alternative but equivalent emissions reduction units should be allowed. On the side of developing countries targeted by the measures, they should utilize the forums provided under the UNFCCC and Kyoto protocol to effectively deal with the measures to their advantage.

RESPONSE OF KOREA TO CLIMATE CHANGE

The final three chapters consider the issues of climate change and mitigation from the standpoint of Korea. The discussion about climate change began in earnest in Korea after the government announced its Green Growth National Strategy. Many policy options have been reviewed under the framework of the strategy, and some of them have already been implemented. There are, however, many remaining challenges. Some people still raise the question of why Korea should take active mitigation action, in a situation where Korea does not belong to the developed country group in the UNFCCC and where the future of international negotiations is uncertain. In terms of policy design, there are concerns whether market based policy tools adopted in developed countries would work properly in Korea.

In Chapter 7, Chin Hee Hahn and Sung-Hyun Ryu approach the need for the green growth strategy from a new angle: an economic viewpoint rather than an environmental one. Their attention is on the decline in terms of trade in Korea since the mid-1990s. They examine the causes of the decline using the decomposition method and regression analyses.

Their decomposition exercise finds that Korea's terms of trade have declined mainly owing to the goods price effect resulting from the rise of oil prices relative to prices of manufactured products. This phenomenon is commonly observed in many countries since the mid-1990s. In addition, their regression results also suggest that China's trade expansion contributed to the decline by raising the prices of oil and raw materials, while lowering the prices of manufactured products.

The empirical findings in this chapter provide some support for the green growth strategy. Korea is well known for its highly energy-dependent economic structure. Most of its energy, moreover, is imported. Taking this into consideration, the findings of the chapter imply that Korea's terms-of-trade decline might persist if economic growth continues in developing countries, especially China, and if Korea clings to its current economic structure. Thus, Korea needs a policy framework to lower its energy dependency, to reduce the external dependency of the economy, and to differentiate its export products from those of China. Innovative policies under the green growth strategy may therefore bring additional benefits by helping to change the economic structure.

Chapter 8 by Jin-Gyu Oh outlines and evaluates the policy framework in the initial stage of Korea's low-carbon green growth strategy, focusing on energy policy. He first examines the current state of affairs in terms of the nation's GHG inventory and energy-related CO_2 emissions in detail. According to the data, the energy sector was by far the most significant emitter, accounting for 84 percent of total GHG emissions. Among emissions from fossil fuel combustion, the electricity generation sector (at 36.1 percent) and the industrial sector (at 31.7 percent) have the largest shares. He then presents the long-term outlook for energy consumption and CO_2 emissions up to 2030, with projections of increases in both. Energy intensity (energy consumption per unit of GDP), carbon intensity (CO_2 emissions per unit of GDP), and CO_2 emissions per unit of energy consumption are projected to decline, implying that energy efficiency and carbon efficiency are expected to improve. The results appear to come mainly from the assumption that the shares of energy-intensive sectors in manufacturing will decline.

The analysis in Chapter 8 shows that the trend in Korea is toward a lower carbon economy. But it is obvious that more efforts are necessary for the mitigation of climate change. Oh identifies two tasks for energy policy: the first is to lower carbon intensity by shifting the energy mix, and the second is to lower energy intensity by improving energy efficiency. The Korean government has already adopted, or is planning to adopt, policies for those purposes, including policies for developing green technology, energy efficiency programs, demand side management programs, promotion of energy efficient vehicles, and policies for nuclear power generation and renewable sources of energy.

The author points out, however, that there are many challenges to achieving these goals. First, in the shift of the energy mix, it will be necessary to gain public acceptance of nuclear power and technology development in the renewable energy sector. Second, current policies for the improvement of energy efficiency rely largely on regulatory measures rather than market based measures. In addition, current price regulation, which keeps prices lower than their costs, hinders efforts to promote more efficient energy consumption. The author recommends greater utilization of the price mechanism and reconsideration of energy pricing policies.

In Chapter 9, Kyoung-Soo Yoon and Min-Kyu Song discuss mitigation policy tools in Korea, in particular the ETS, beginning with the issue of policy mix. They argue that market based policies are superior, by inducing an efficient balance of input-substitution cost, reduction of abatement cost, and output-substitution cost. However, they also note that it may be inevitable, in the actual reduction policies, to combine multiple policy instruments, owing to multiple market failure factors and political constraints. But even in this case, it will be necessary to closely monitor the interaction between policy instruments, as a means of avoiding double regulation problems and also ensuring a comprehensive policy in order to prevent carbon leakage. As illustrated in Chapter 8 in detail, many policies are proposed for similar policy objectives in Korea. For this reason, designing an efficient set of policy instruments will be critical to minimize the mitigation cost.

The chapter then addresses a few issues about the design of the ETS in Korea. First, concerning the allocation of tradeable permits, the authors argue that auction allocation is superior in terms of allocative efficiency, macroeconomic efficiency, and equity. However, considering the compensation for the decline in corporate value and for the policy acceptability during the initial adoption period, a gradual increase of the portion of auctioning from mostly free allocation may be desirable. Second, special attention should be paid to the power sector in Korea, because of the high market concentration, the cost based pool system in the wholesale market, and price regulation for electricity. It is probable that ETS in the power sector will not work properly because the emissions cost will not be efficiently passed through to the consumer price, especially when the free permit allocation system is adopted. To avoid such a problem, the government might consider applying a carbon tax to the power sector, or adopting a pricing scheme that links the consumer price to the permit price.

The chapter then discusses the design of a permit trading market. Unless the Korean carbon market is connected directly to outside big markets such as the EU ETS, it is expected that the market will be very thin, especially in the initial period. Under these circumstances, the market should be designed for efficient price discovery and provided with risk management tools, to keep the volatility of prices low. Taking these factors into account, the authors emphasize that balances between transparency and anonymity and between the exchange and the over-the-counter markets should be pursued. They further recommend predictable reduction targets over time, active derivatives markets, and possible additions of a central clearing house (the central counter party or CCP system) and carbon central banking.

Climate change is a global, long-lasting phenomenon that the world must cope with in the twenty-first century. The authors in this book assert in common that responding to it is inevitable, but also acknowledge that there are many difficulties and large uncertainties. During the past two decades, many countries have made great strides toward the goal of mitigating climate change, but these were just the first steps in a long journey. Much more experience is needed in order to develop policies that are cost effective and acceptable, especially for developing countries and newly developed countries such as Korea. To do so and to design creative policies based on sound principles, valuable lessons can be drawn from the limited experiences already available. This book examines such lessons now emerging from recent experiments and raises key issues from the perspective of Korea, as one among many newly developed nations that will contribute to this global effort.

REFERENCES

International Energy Agency (IEA). 2009. *CO$_2$ Emissions from Fuel Combustion Highlights.* Paris: International Energy Agency.

National Institute of Meteorological Research (NIMR). 2009. *Understanding the Climate Change II: Climate Change in the Korean Peninsula at the Present and in the Future.* Seoul: National Institute of Meteorological Research. In Korean.

Organisation for Economic Co-operation and Development (OECD). 2009. *Climate Change Mitigation: Policies and Options for Global Action beyond 2012.* Paris: Organisation for Economic Co-operation and Development.

Rodrik, Dani. 2005. "Growth Strategies." In *Handbook of Economic Growth,* edited by Philippe Aghion and Steven N. Durlauf, pp. 967–1014. Amsterdam: North Holland.

Stern, Nicholas. 2007. *The Economics of Climate Change: The Stern Review.* Cambridge and New York: Cambridge University Press.

World Bank. 2008. *The Growth Report: Strategies for Sustained Growth and Inclusive Development.* Commission on Growth and Development. Washington: World Bank.

2. Sustaining growth and mitigating climate change: are the costs of mitigation underestimated?

Stephen Howes

INTRODUCTION

Despite many estimates showing that the costs of mitigation of climate change will be moderate, concerns remain that mitigation will slow and possibly halt economic growth. Take for example the conclusion of Dieter Helm, who, though he supports the "case for urgent action" (Helm 2008:236), argues:

> The happy political message that we can deal with climate change without affecting our standard of living—which is a key implicit message from the Stern Report on which politicians have publicly focused—and do so in a sustainable way, turns out, unfortunately, to be wrong. (Helm 2008:228)

This chapter explores the concerns of those such as Helm in relation to standard and often-used estimates of mitigation costs, and assesses their validity. It begins with a brief survey of mitigation cost estimates, and then outlines and considers a range of issues in relation to them. The aim is to address in particular the apprehensions around whether the standard estimates of mitigation costs are too low. Whereas Weyant (2000) and Barker et al. (2006) explore reasons why different models give different cost estimates, the concern in the discussion below is more with cost considerations that might not be captured by the models from which global cost estimates are derived. The chapter draws in particular on the recent Australian Garnaut Review of Climate Change (Garnaut 2008), the related modelling of the Australian Treasury (2008), and the public debate in Australia in relation to the introduction of an emissions trading scheme. The chapter concludes that the national costs of mitigation may well lie above the commonly estimated range of, say, 1–5 percent of GDP. The two most serious risks are that in some countries the national costs of mitigation will be significantly above the global average—some modelling

Table 2.1   Survey of global mitigation cost estimates for 2050

Modelling exercise	Mitigation cost in 2050 as a percentage of global output
550 ppm CO$_2$eq stabilization target	
IPCC Fourth Assessment Report (2007a) 535–590	1.3 (slightly negative to 4)
Stern 550 (2007)	1 (−2 to 4)
Australian Treasury GTEM 550 (2008)	2.7
450–500 ppm CO$_2$eq stabilization target	
IPCC Fourth Assessment Report (2007a) 445–535	(up to 5.5)
Stern 500 (2009)	2
Australian Treasury GTEM 450 (2008)	4.3

Notes: Costs are mean or individual estimates (with ranges in parentheses). Note that the 535–590 ppm CO$_2$eq range is assessed to hold on the basis of "high agreement, much evidence" (IPCC 2007a:172). The 445–535 ppm CO$_2$eq range is assessed to hold with "high agreement, medium evidence." There are fewer studies with this more stringent stabilization target, which is also why only a range is presented and no mean.

suggests that this is particularly a risk for poor countries—and that policy mistakes will drive up the costs of mitigation. These factors are unlikely to make mitigation not worth undertaking, but could discourage mitigation effort. Policies that are efficient and that assist developing countries to meet some of their mitigation costs will likely have a high environmental return.

Table 2.1 provides a survey of global mitigation costs as a percentage of global output for the year 2050 and for two stabilization targets: about 550 parts per million (ppm) of carbon dioxide equivalent and a more stringent target of about 450 ppm. The table shows the range of results reported in the Fourth Assessment Report of the International Panel on Climate Change (IPCC), the cost estimates from Stern (2007 and 2009, probably the most influential in the global debate), and some recent estimates derived by the Australian Treasury (2008), using the Global Trade and Environment Model (GTEM), which are drawn on later in the chapter.

The best known of these cost estimates is Stern's (2007) 1 percent of global output. Stern (2009) now supports a 2 percent cost estimate to achieve a tougher mitigation target (500 rather than 550 parts per million). Stern's estimates are at the low end of the range of published estimates. The recently published Australian Treasury GTEM results put the cost of global mitigation at twice or more Stern's levels. The Stern review uses a bottom-up model based on cost estimates of different technologies. The Treasury and other general equilibrium modelling exercises derive their estimates from models that have less technological detail but capture second-round effects such as reductions in savings (caused by income losses) and thus declines in investment. Weyant (2000) concludes from his survey that the most important determinants of modelled mitigation costs are the size of the abatement task relative to GDP, and the scope for flexibility allowed in the policy regime. Differences in assumptions about the flexibility of the economy to substitute and about the speed with which new technology develops have a secondary influence. Barker et al. (2006) reach similar conclusions.

In absolute terms, the cost estimates in Table 2.1 are large. Even 1 percent of world output in 2050—measured in 2005 U.S. dollar prices, using purchasing power parities to compare across countries— is US$2.7 trillion or about the size of the Indian economy today. Nevertheless, costs in the range presented in Table 2.1 would have only a minor impact on growth. Their growth impact can be measured (1) as a growth penalty (the difference between annual average growth under the business as usual or no-mitigation scenario and under the mitigation scenario) or (2) as a growth delay (the additional time it will take the world to achieve the no-mitigation level of output). Figure 2.1 shows how the cost of global output translates into growth penalties and growth delays. It illustrates that even relatively large costs—if incurred over a long period of time and in the context of reasonable underlying growth—have only a limited impact on growth rates and impose only moderate delays in achieving given growth targets. Thus, for example, the Stern 1 percent 2050 cost estimate translates into a growth penalty of 0.02 percentage points and to a growth delay of 4 months. Even the Australian Treasury (450 ppm) 4 percent cost estimate amounts to a growth penalty of just under 0.1 percentage points and a growth delay of 17 months.

From Figure 2.1, the growth impact of even a stringent mitigation target costed using more conservative assumptions than Stern appears to be minor. Why then is the cost of mitigation still a contentious issue? There are several possible reasons. First and most important, even if the mitigation cost estimates presented above are accurate, and even if they amount to only a minor tax on growth, they might still not be worth incurring. Second, decisions to mitigate are made at the national not the international level. Such decisions will be based not on global but on national mitigation cost estimates, and these might range widely around the global average. Third and fourth, the standard global cost estimates typically include a number of unrealistic assumptions that push costs

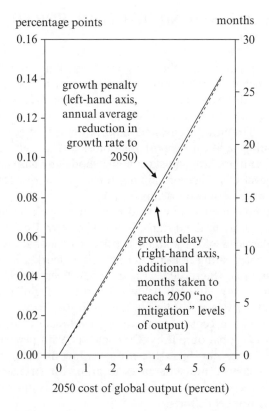

percentage points   months

Note: The assumed annual average global growth (without mitigation) from 2005 to 2050 is 3 percent.

Figure 2.1   Growth penalties and growth delays for different 2050 mitigation costs

down, including, respectively, optimal domestic policies and complete global participation.[1] Fifth, modelling exercises also typically neglect short-term transitional costs and the distribution of costs. Finally, there are also counterarguments that mitigation cost estimates are systematically overstated and ignore the potential benefits of climate change policies in driving economic innovation and growth.

The following sections interrogate the standard mitigation cost estimates with respect to these various concerns. As noted earlier, the aim is not to provide an explanation of why different published global cost estimates differ, but rather to consider whether all the estimates might be biased downward (or perhaps upward).

ARE THE COSTS OF MITIGATION WORTH INCURRING?

Frankel (2005) notes that the economic modelling in which he was involved, on behalf of the Clinton administration in the United States, showed that, with heavy reliance on trade in permits, it would cost the United States only 0.1 percent of GDP a year to fulfil Kyoto Protocol commitments. The clear, if unstated, implication is that even this level of costs was testing political support for domestic U.S. mitigation at that time. Indeed, until recently, most economic modelling has supported only a moderate level of mitigation, aiming for targets above 550 ppm (Toll 2006). There have been two reasons for this.

First, economic models have tended to show high levels of climate change damage only in the distant future. The Stern Review gave a best estimate of damages of only about 3 percent of global output in 2100 even when incorporating non-market as well as market impacts as well as catastrophic risk (Stern 2007: Figure 6.5c). Even the 95th percentile damage estimate for 2100 fell below 10 percent of global output. The costs of climate change according to the Stern Review were much larger in the twenty-second century, with the best estimate of damages rising to 14 percent of global output by 2200, and the 95th percentile estimate increasing to 35 percent. Since Stern used a low discount rate, he used these twenty-second century damages to argue for stringent mitigation today, but most economists worked with higher discount rates and thus supported less stringent efforts.

The debate about discounting continues, but more recent projections of emissions under business as usual bring forward the damages of climate change. Stern assumed a temperature increase over the course of the century of about 3.4 degrees Celsius, right in the middle of the IPCC Fourth Assessment Report central range of 2.3 to 4.5 degrees.[2] The Garnaut Review (2008), by contrast, built its no-mitigation emissions projections on the assumption of long-term, rapid developing-country growth. As a result, it derived a business-as-usual temperature increase projection this century of 5.1 degrees Celsius, which is above the IPCC range.

Second, many economic models show relatively limited impacts of climate change, even for high temperature increases. Consider a 5 degree Celsius increase. This is above the range of "tipping points" for seven of the eight catastrophic global events, for which Lenton et al. (2008) present such a range, and at the top end of the range for the eighth.[3] A recent study by the Center for Strategic and International Studies concluded that this extent of climate change would pose almost inconceivable challenges as human society struggled to adapt. . . . The collapse and chaos associated with extreme climate change futures would destabilize virtually every aspect of modern life. (CSIS 2007:7, 9)

However, the survey of economic models in the IPCC Fourth Assessment Report (2007b) suggests that, for this level of temperature increase, global output would take a hit of somewhere between about 0 and 10 percent (see Figure 20.3 in Chapter 20 of IPCC 2007b): significant, but far from catastrophic, and indeed as Figure 2.1 shows, quite manageable in the context of a century of growth.

Worryingly, this disconnect may be increasing over time. Smith et al. (2009: Figure 1), summarizes how much more risk scientists have come to attach to even moderate levels of climate change between the third (2001) and fourth (2007) IPCC assessment reports. But the survey of aggregate economic costs of climate change given in the IPCC third and fourth assessment reports shows no sign of an upward trend in the relationship between temperature and damage, as is perceived by economists.[4]

A full account of this disconnect is beyond the scope of this chapter. However, it is important to recognize the limits of economic modelling when it comes to estimating climate change impact. Impacts might be catastrophic, and yet be difficult to model simply because they are difficult to quantify. It is obvious that an unmitigated future will be hugely damaging and risky. In the words of Martin Weitzman, a large temperature increase, of the sort that would occur if emissions continue to grow rapidly (Weitzman refers to an average increase of 6 degrees Celsius), would result in

a terra incognita biosphere . . . whose mass species extinctions, radical alterations of natural environments, and other extreme outdoor consequences of a different planet will have been triggered by a geologically instantaneous temperature change that is significantly larger than what separates us now from past ice ages. (Weitzman 2007:717)

The question then is not whether to mitigate but how much. Here again, the role of modelling might be limited. Modelling carried out by the Garnaut Review suggests that the climate change associated with 450 and 550 ppm levels of mitigation will give rise to roughly similar expected market impacts. Figure 2.2 illustrates by showing not the costs of different levels of mitigation but the costs associated with the climate change associated with these levels of mitigation. There is very little difference between the modelled damage from stabilization at 450 ppm and 500 ppm. Given this, it would only make sense to opt for more stringent (and

cost as a percentage of GNP,
relative to the reference case

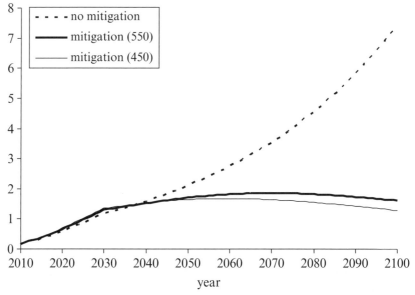

Note: These estimates are achieved by "shocking" the reference (no climate change) case with the differing levels of impact associated with the temperatures expected from the three scenarios. There are no costs of mitigation.

Source: Garnaut (2008).

Figure 2.2   Expected, modeled market costs for Australia of unmitigated and mitigated climate change, 2010–2100

therefore more expensive) mitigation for the greater protection it would give against non-market damages and against low probability catastrophic events, both of which are hard to model. The Garnaut Review argued on these (non-quantified) grounds that Australia should support 450 ppm as a global stabilization objective even though this would cost about 1 percent of GNP more (in net present value terms) over the course of the century.

In summary, even if it is not evident from climate change damage estimates based on economic modelling, the sorts of mitigation costs presented in Table 2.1 are certainly modest enough to warrant even stringent levels of mitigation. Whether they can be relied on is the subject of the rest of this chapter.

NATIONAL MITIGATION COSTS

Decisions to mitigate are made not at the international but at the national level. From this perspective, the global mitigation cost estimates surveyed in Table 2.1 are of limited interest. It is well known that climate change damages will vary from region to region, with cold areas even benefiting from limited amounts of global warming. Mitigation costs will also vary from region to region. But modelling exercises typically pay less attention to national than international mitigation costs.

How global mitigation costs are shared around the world depends on burden sharing arrangements. The global mitigation costs presented above assume efficient mitigation, with a single global price on greenhouse gas emissions. A useful benchmark to consider is the national cost of the mitigation that would transpire if each country was subject to the same global carbon price, prior to any international transfers.

Figure 2.3 shows the great variation in mitigation costs across countries, according to the GTEM model used by the Australian Treasury (2008).

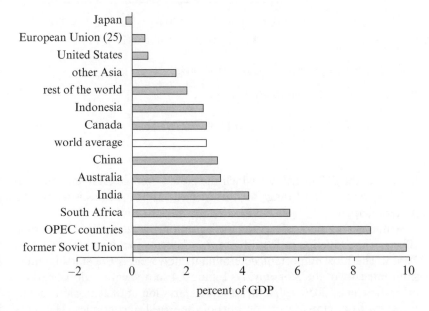

Note: Results were generated using the GTEM model.

Source: Australian Treasury (2008), Table 5.14.

Figure 2.3 GDP costs of mitigation for 2050 under a 550 ppm global mitigation strategy: selected countries and regions

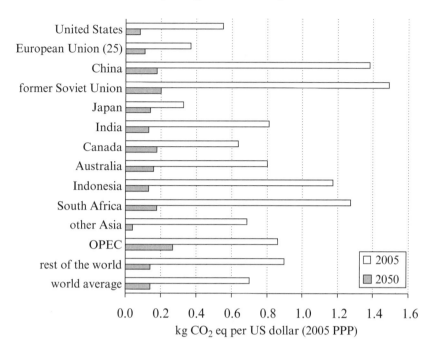

Note: Results were generated using the GTEM model.

Source: Australian Treasury (2008), Table 5.11.

Figure 2.4   Emissions intensity in 2005 and 2050 under a 550 ppm global mitigation strategy

If we take the 550 target, for which the global average GDP cost in 2050 is 2.8 percent, the cost range for the thirteen GTEM regions is from −0.2 percent (Japan) to 9.9 percent (the former Soviet Union).

A number of factors explain the variation, including terms of trade effects. The reduction in fossil fuel prices benefits fuel importers, such as Japan. The most important determinant, however, is the starting emissions intensity of the economy. As Figure 2.4 shows, with a global carbon price there is by 2050 very little absolute variation in the emissions intensities (ratios of emissions to output) of the world's economies. The units are kg of CO_2 equivalent per U.S. dollar (again using purchasing power parities at 2005 prices). The 2050 range is from 0.04 (other Asia) to 0.27 (OPEC). Compare that to the range today (2005), from 0.33 (Japan) to 1.49 (the former Soviet Union).

At least according to GTEM, a common carbon price will induce all

percent of GDP for 550 global mitigation
(2050 cost of mitigation)

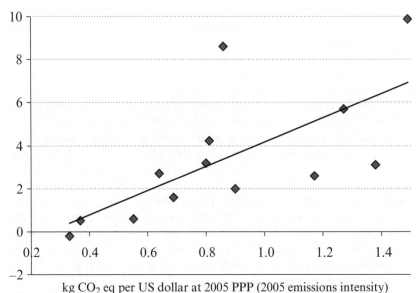

kg CO_2 eq per US dollar at 2005 PPP (2005 emissions intensity)

Note: The results were generated using the GTEM model; see Figures 2.3 and 2.4.

Source: Australian Treasury (2008).

Figure 2.5 Cost of mitigation in 2050 and emissions intensity in 2005

countries to move not only in the same direction but to more or less the
same destination emissions intensity. Those with the longest to go to reach
this destination incur the highest costs. Figure 2.5 shows the close correlation between initial emissions intensity and the 2050 cost of mitigation.

One striking consequence of this close correlation is that on average
developing countries incur the highest mitigation costs. GTEM 2050
global costs for the 550 mitigation strategy, which are 2.8 percent of global
output using purchasing power parities, are only 2.2 percent if market
exchange rates (which give developing economies a lower weight) are used
to aggregate costs across countries. China, India, Indonesia, South Africa,
OPEC, and the rest of the world (essentially Latin America and Africa)
all have above average initial emission intensities today (Figure 2.4). They
also all have above average 2050 mitigation costs, except for Indonesia
and the rest-of-the-world category, which evidently find it cheap to reduce
deforestation (Figure 2.3).

It remains to be seen whether this result holds across models. The other global model used by the Treasury, G-cubed, also shows significantly higher costs of global mitigation when assessed using purchasing power parities rather than market exchange rates (2.5 percent versus 1.9 percent for the 550 strategy).[5]

This result may appear to be contrary to the received wisdom that mitigation will be cheap in developing countries. For example, Frankel (2005:46) writes: "It is far easier for some countries to cut emissions relative to the BAU [business-as-usual] path than for others." He gives China and the United States as examples of places where it will be, respectively, cheap and expensive to reduce emissions. Many developing countries have low marginal mitigation costs, but this may simply mean that they mitigate a lot, and so still end up with high total mitigation costs.[6] The modelling results reported here do not include the costs of stranded investments, for example mothballed coal-fired power plants. If they were included, the costs of developed country mitigation could go up relative to those in developing countries since the former could have more stranded assets (Stern 2007:275; Frankel 2005). This notwithstanding, these results should clearly caution us against simplistic statements that mitigation will cost developing countries less.

The results also underline the importance of developed countries putting in place generous transfer programs to help compensate developing countries for the costs they will face. This will reduce costs for developing countries, and increase them for developed countries. This in turn will reduce the dispersion in mitigation costs between developing and developed countries, and in general push countries toward the global average.

POOR POLICIES

The cost estimates provided in Table 2.1 assume that the cheapest abatement opportunities are selected. In bottom-up models, mitigation options are simply ranked from cheapest to most expensive, and selected accordingly. In top-down models, a single instrument, a carbon price, ensures that the cheapest abatement opportunities are selected first. The real world is more complex. In practice, countries use a range of policies, in addition to or in place of carbon pricing, including separate targets for renewable energy and for biofuels, government R&D programs, and often separate targets or policies for energy efficiency or product quality regulation (for example automobile efficiency standards).

Both political and economic arguments can be made for this portfolio approach to mitigation policy (Hannemann 2007, 2008). In an environment of political uncertainty, it might make sense to use a diversified approach. For example, a government might not be sure if it can introduce carbon pricing or it might be worried that if carbon price rises too high, political support will be compromised. Such governments will have an incentive to seek overlapping policies in case carbon pricing is unavailable as a policy tool, or to dampen carbon prices.

From an economic perspective, as Bennear and Stavins (2007:111) argue, "under a fairly broad set of circumstances the use of multiple policy instruments can be justified as optimal in a second-best world." For example, while cap and trade worked to reduce sulfur emissions from coal-fired power plants, in that case the technology to respond to the problem at hand was well developed. Some of the technology required for successful mitigation is still under development, and a range of policies may be needed to induce that technology (Grubb 2004).[7] In addition, mitigation can give rise to national first-mover advantages, if, as discussed in the penultimate section, particular countries are the first to display mastery of particular technologies. However, this requires not a dispersed market-based effort, but a focus on particular technologies, supported by a variety of instruments (see the example of Denmark and wind power discussed below).

While these arguments for the use of multiple mitigation policies have some merit, it is also the case that the greater the variety of policies used, the greater the scope for political rent-seeking and the greater the likelihood that really bad policies will be adopted. Policies relating to biofuels are the most egregious example to date. One study of OECD biofuels estimated "costs ranging from US$150 to over US$1,500 per metric tonne of CO_2-equivalent avoided" (Kutas et al. 2007). Other studies have questioned whether biofuel use actually reduces greenhouse gas emissions, noting that some methods of producing biofuels "actually increase global warming due to land conversion and the release of huge amounts of carbon that otherwise would remain in plants and soil" (CGIAR 2008). Biofuel policies have been used to subsidize domestic producers and keep out more efficient foreign producers. The United States imposes a tariff of US$0.54 per gallon on ethanol imports to keep Brazil, the world's lowest-cost producer, out of its market (Smith 2007). The rapid growth of the ethanol industry has pushed up food prices, hurting the poor in developing countries (CGIAR 2008).

Renewable energy mandates are also expensive. A study by the U.K. National Audit Office (2005) found that the United Kingdom's renewable obligation policy (which mandated 10 percent of U.K. electricity to come from renewable sources by 2010) saves greenhouse gas emissions at the high cost of 70–140 euros per tonne of CO_2. The German policy of promoting domestic photovoltaic solar has been criticized as being particularly expensive. The generous feed-in tariff that has supported the expansion of solar photovoltaics in Germany is now being wound back owing to cost pressures.[8]

Policy uncertainty can be another source of cost. Investors will delay decisions if it is unclear which energy policies will be implemented when, and with what vigor. At the extreme, this could compromise energy security and reliability.

Finally, as noted in the next section, the adoption of measures to shield emissions-intensive industries can give rise to rent seeking, as industries clamber to be on the list of protected industries. It could also give rise to trade disputes, if countries seek to penalize those seen to be obtaining an unfair trade advantage through not mitigating.

It is difficult if not impossible to quantify the policy costs likely to be associated with climate change mitigation. Dieter Helm argues that on account of policy costs, among other factors, the "costs of mitigating climate change are likely to be significantly higher than indicated by the Stern Report" (Helm 2008:228). Ross Garnaut (2008:297) estimates that, if handled incorrectly, the special treatment of emissions-intensive trade-exposed industries, discussed further in the next section, could "turn out to be as expensive as the costs of mitigation itself." Nicholas Stern's most recent book emphasizes "the importance of good policy in keeping costs down" (Stern 2009:55).

INCOMPLETE PARTICIPATION

A critical concern for countries considering whether to mitigate is incomplete participation. In the debate in Australia over the introduction of an emissions trading scheme, no issue has received more attention. A speech by the opposition Liberal Party spokesman illustrates the arguments against "going it alone."

> . . . global emissions could actually increase as investments and jobs, especially from major regional centres, leave Australia and go to developing countries where less efficient factories pump out much more CO_2 than in Australia. And without our major competitors engaging in some form of scheme the cost to Australians will be much greater. This cost will be measured in the premature closure of many coal mines, cement works, coal powered generators and fuel refineries and the loss of major investment in new smelters, metal refineries, LNG gas projects, cement works, exploration and much more. (Robb 2009)

A world in which only some countries decide to mitigate makes mitigation less attractive for the mitigating countries for three reasons. First, with incomplete participation, participating countries have to do more to achieve any given global environmental target (Nordhaus 2008). Nordhaus calculates that a participation rate of 50 percent imposes a cost penalty of 250 percent for this reason.

Second, costs for countries subject to national emissions caps, like those in the Kyoto protocol, will rise if the opportunities for trade in permits are limited. This phenomenon has been extensively modelled. Some estimates show costs being halved or more by the introduction of trading—relative to a global mitigation program in which there are national caps but no trades (Stern 2009:164). Brandt and Svendsen (2006: Table 2) present a range of estimates which show that the cost for the United States of achieving its Kyoto target was three to ten times more expensive without trading than with.

Third, and this is the aspect that causes the greatest concern, incomplete participation gives rise to the risk of leakage, that is, an increase in emissions in non-participating countries.[9] This leakage can happen either directly (as emissions-intensive production moves to non-participating countries) or indirectly (as fossil fuel prices fall in non-participating countries owing to reduced global demand). The global environmental benefits of domestic mitigation are thus reduced. It is also often claimed that the unlevel playing field (a result of incomplete participation) will in itself increase the costs of domestic mitigation, but this is not necessarily so. McKibbin and Pearce (2007) consider a carbon tax, and so abstract from the impact of participation on international permit trading. They analyze a carbon tax implemented only in Australia, only in the rest of the world, and then all over the world. The results show that mitigation is actually cheaper in Australia when it is going alone (see McKibbin and Pearce 2007:26, Figure 3.1). This is because of the terms-of-trade effect that international mitigation has: demand from the rest of the world for Australian coal in particular falls. Thus, incomplete participation does not necessarily make domestic mitigation more expensive, but it does make it environmentally less effective.

The likely extent of leakage is very uncertain. The available evidence is surveyed in Chapter 11 (section 11.7) of the report of the Third Working Group of the IPCC's Fourth Assessment Report. This cites a leakage range of 5–20 percent (the ratio of increase in emissions in non-participating countries to decrease in participating countries), but there is little evidence on which to base such a claim, since the so-called participating countries have either not implemented economywide policies or have shielded their trade-exposed emissions-intensive industries. Many analysts suggest that direct leakage will be small, citing experience with other environmental regulations (Aldy and Pizer 2009; Stern 2007: Chapter 11), and the fact that emissions-intensive trade-exposed industries are only a small proportion of total output. In Australia, the ten industries with the highest emissions per unit of revenue in 2001–02 contributed 37 percent of national emissions, 4 percent of national production, 3 percent of employment, and 15 percent of exports (Australian Government 2008:313). The fact that emissions-intensive industries are also capital intensive suggests that relocation is unlikely. Firms establishing new plants will look at many factors; with respect to carbon prices, the prospect of carbon price introduction in the future will deter the establishment of long-lived carbon-intensive projects in countries that currently do not have carbon prices.

Even if direct leakage is an exaggerated and misunderstood problem in the public debate, it is politically salient. In a world with incomplete participation, experience shows that countries will not mitigate without providing some protection to their carbon-intensive, trade-exposed industries. This protection will typically not only involve compensation (which would have no impact on aggregate costs) but also some sort of shielding, that is, the full or partial exemption of the industries concerned from the carbon pricing regime. Shielding will tend to increase costs since it reduces effective coverage. Modelling for the Garnaut Review suggested an additional cost impact of shielding of about 0.2 percent of GNP by 2020 (1.7 percent of GNP cost with shielding versus 1.5 percent without) rising to as much as 0.5 percent of GNP by 2030 (4 percent versus 3.5 percent).[10]

Additional costs from leakage can also arise as a result of rent seeking associated with the policy response, or even potentially as a result of trade disputes. This is a form of the policy costs discussed in the previous section. Ross Garnaut has noted that the "arbitrary nature" of assistance measures to address leakage concerns "will make them the subject of intense lobbying with potential for serious distortion of policy-making processes" (Garnaut 2008:317) and "has the capacity to . . . pervert individual domestic schemes to the point of non-viability" (Garnaut 2008:342).

Finally, experience suggests that incomplete participation will result in participating countries doing less. Incomplete participation discourages action by participating nations, in part because of worries about carbon leakage and competitiveness, and in part because of a perception that climate change mitigation, if limited to a few countries, is both unfair and ineffective. The decision of the "enthusiastic countries" (to use the terminology of Victor 2008) to continue to mitigate, albeit at a lower level, despite incomplete participation is made in response to public pressure, in the hope of encouraging the "reluctant countries" to do more over time, to enable a gradual decarbonization of the economy and avoid sharp adjustments and, as discussed in the next-to-last section below, to gain first-mover advantages.

Both the European Union and Australia have formally made conditional and unconditional offers. The European Union has said it will reduce emissions by 20 percent over 1990 levels by 2020 unilaterally, and by 30 percent if there is a global agreement. Australia has said it will reduce emissions by 5 percent by 2020 over 1990 levels unilaterally, and by 15 percent or 25 percent if there is a global agreement (depending on the strength of that agreement). The Garnaut Review estimated that the 2020 GNP cost of the unconditional offer was 1.4 percent, little different from the cost of the 15 percent target (1.5 percent) or the 20 percent one (2.0 percent). The greater ambition of the conditional offers is almost fully offset by the greater potential for international trade in permits afforded by an international agreement (Garnaut 2008: Table 12.3).[11]

This reaction from participating countries (to do less) will prevent incomplete participation leading to an escalation of domestic mitigation costs. It will of course do nothing to help achieve the environmental objectives of mitigation.

TRANSITIONAL AND DISTRIBUTIONAL COSTS

Policymakers are not only, or even primarily, concerned with the aggregate, long-run costs reported at the start of this chapter. They are also concerned with the costs that will fall in the near term and on particular regions or industries or on groups of households or workers. In the long run, resources will adjust, and there will be both losses in and gains to employment. But political concerns are likely to focus more on existing jobs, and threats to them, particularly when they are in concentrated geographical areas.

Though some mitigation cost models incorporate adjustment costs and so allow for rises of unemployment, in general these models are not a strong basis on which to base short-term forecasts. Figure 2.6 illustrates this point with the two models used in the Garnaut Review (2008). It shows projected emissions (relative to their 2012 level) from the introduction of a carbon price into Australia in 2013. Though by 2030, the two models show a similar level for domestic emissions, the two models show very different emissions paths prior to 2020. They also both show implausibly large single-year falls in the year in which the carbon price is introduced: in GTEM, emissions fall by 14 percent in a single year, and in MMRF by 9 percent. An emissions decline of this magnitude in a single year is unlikely, and one would place more faith in the convergent 2020 results than in the divergent 2013 ones. The introduction of adjustment

2012 emissions = 1.00

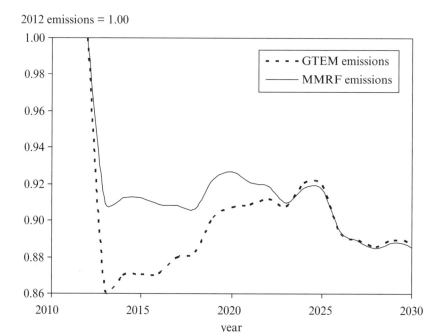

Note: A carbon price consistent with a 550 ppm stabilization path is introduced in 2013.

Source: Modeling undertaken for the Garnaut Review (2008).

Figure 2.6   Emissions in Australia 2010–30 following the introduction of a carbon price in 2013, according to two different models used by the Garnaut Review

costs would yield more plausible and smaller short-run changes, but given our limited experience with carbon pricing, it is difficult to know just how costly the adjustment will be.

Similarly, some modelling work incorporates distributional costs, but regional estimates, which are often the focus of policy concern, are particularly fraught with uncertainty.

While it is difficult to model the short-term, transitional and regional impacts of mitigation, policymakers do have tools available to manage these impacts, namely through programs of structural adjustment, similar to those often used in response to programs of tariff reform.

Another form of transitional costs that might be incurred is the rise in prices due to mitigation-induced demand surges, for example, for renewable technologies. Helm notes that the prices of wind turbines have risen sharply as the dash-for-wind has been embedded in renewables policy; and now there is evidence in the sharply rising prices of new nuclear development technologies as manufacturing production lags demand. These price effects are of a significant order of magnitude. . . . (Helm 2008:225)

The presence of these rents does not increase economic costs (since they are in the nature of a resource transfer), but they will increase the costs to households (for example, resulting in higher electricity prices) and thus make the transition politically more difficult.

A MITIGATION STIMULUS?

The previous sections have examined whether the standard estimates of mitigation costs might be systematically biased downward. Are there any reasons why they might be systematically biased upward?

There is some evidence that past predictions of environmental compliance have been overly pessimistic. Hodges (1997) examines twelve cases of new pollution regulations, and finds that in eleven of them early cost estimates were more than double later estimates (and for the twelfth, was 30 percent above). Will climate change be another example where we have underestimated the human capacity for innovation? We do not know. But it is important to note that the current range of models already reflects a range of views on technological development. Climate change mitigation will demand a restructuring of the economy, and it is unclear whether one can extrapolate from the experience of dealing with other, far less costly environmental problems.

Some point to broader, spillover impacts from mitigation spending. In the short term, such spending could help boost demand during a slow-down, an argument that became popular in the response to the global financial crisis. In the words of Gordon Brown, Prime Minister of Great Britain (2009): "the drive to a low carbon economy is not something to be delayed because of the global recession; instead it can be a powerful driver of global recovery." In the long run, however, mitigation spending is an unlikely counter-cyclical fiscal policy tool.

The longer-term argument that mitigation will drive the next round of innovation and growth has also become increasingly popular. Nicholas Stern (2009) makes this case.

The changes in technology required to get to a low carbon world are likely to usher in a burst of innovation, creativity and investment. . . . The new technologies and investment opportunities of low carbon growth will be the main drivers of sustainable growth in the coming few decades. These investments will play the role of the railways, electricity, the motor car and information technology in earlier periods of economic history. (Stern 2007:47, 206)

The 2009 *World Economic and Social Survey* argues that the large-scale investments needed to respond to climate change could trigger "virtuous growth circles" in developing countries (United Nations 2009:xv).

Whether mitigation will not only drive but also accelerate global growth at the global level will depend on whether it will stimulate additional innovation and investment, or rather shift existing innovation and investment from other areas (see Weyant 2000; Jochem and Madlener 2003). This seems uncertain, and difficult to measure. It is more straightforward that individual countries could benefit from early mitigation by obtaining a first-mover advantage in particular mitigation technologies. This is essentially an infant industry argument, best illustrated by Denmark's success in the wind-turbine industry, not only domestically but subsequently overseas (Brandt and Svendsen 2006). Denmark has about 3,000 megawatts (MW) of installed windpower capacity, but Danish firms have installed about 20,000 MW globally, about 40 percent of the world total. Denmark started promoting wind power in the mid-1970s. One review of the Danish experience concludes that: "The careful balance and timing of R&D and procurement support [including investment subsidies and feed-in tariffs] have both been important to promote both innovation and diffusion of wind energy" (Klaassen et al. 2005:231).

Brazil's experience with sugar-based biofuels provides another example. Almost 20 percent of Brazil's automotive fuel now comes from ethanol, and the industry runs without subsidy, generating significant employment and greenhouse gas savings (provided that the additional demand for sugar cane does not lead to deforestation, which is an open question).

As discussed earlier, the potential for reaping first-mover advantages is an argument in favour of domestic mitigation in the context of incomplete global participation. And this mixing of objectives of industrial policy and climate change policy would lead one to consider policies (such as renewable energy mandates) that one would otherwise regard as inferior to a carbon price. But one's judgement on whether these potential first-mover advantages are worth pursuing will depend on broader views of the importance and appropriateness of countries pursuing activist industrial policies, and the likelihood of governments being able to "pick winners." If one is an optimist in these matters, one will see gaining a first-mover advantage as an upside to incurring mitigation costs; if one is a pessimist, one will see only an additional downside risk to embarking on mitigation, namely that of failed industrial policies.

CONCLUSION

The second section of this chapter argues that if the standard mitigation cost estimates are reliable, then given the risks of unmitigated or even moderately mitigated climate change, there is a strong case for stringent mitigation. The standard cost estimates are typically in the range of 1–5 percent of global output by 2050. Are they reliable? Various criticisms of them can be made. Five are considered in this chapter. The first two emerge as well founded and the other three less so. Working in reverse order through the chapter, I summarize each in turn below.

First, it is possible that the global and national costs of mitigation might be not under- but over-estimated. At the national level there can be first-mover advantages, and at the global level dynamic gains from the investment and innovation that mitigation will no doubt bring. However, these gains appear too uncertain to be given much weight, and the activist industrial policies needed to pursue a first-mover advantage could just as likely add to mitigation costs as reduce them.

Second, the short-term, transitional costs of mitigation are difficult to predict, and the costs of mitigation will be unevenly distributed across society. This makes the case for structural adjustment programs to accompany mitigation, and no doubt increases the political costs of mitigation. However, it does not suggest that aggregate mitigation costs are underestimated, nor does it undermine the case for mitigation.

Third, though incomplete participation is often presented as putting significant upward pressure on mitigation costs, the reality is more complex. Many developed countries will suffer an increase in mitigation costs if opportunities for international trade in permits are limited or made more expensive as a result of incomplete participation. Incomplete participation also reduces the global environmental benefits of domestic mitigation through leakage. But whether or not it increases domestic mitigation costs depends on the feedback from global mitigation to the domestic economy, for example, through demand for exports. The shielding given to emissions-intensive trade-exposed sectors can also be a source of additional costs. Most importantly, in practice, countries have shown that they will reduce levels of mitigation ambition in situations of incomplete participation, and thereby limit domestic costs, albeit at the expense of the environment.

This leaves two problems with the standard mitigation cost estimates, which I conclude are real causes for thinking that the standard estimates are under-estimates.

First, perhaps the biggest risk to mitigation costs comes from poor policies. Recent experience with biofuels and renewable energy suggests that policymakers may choose mitigation options that are far from least cost. Even granting that first-mover advantages might flow from the concerted and systematic pursuit of particular mitigation policies, and that the political and economic challenges of mitigation require a multifaceted policy response, it is hard not to be alarmed by the scope for costly policy error.

Second, decisions to mitigate will be made not at the global but at the national level. Mitigation costs will vary significantly from country to country. This chapter presents evidence to suggest much higher domestic mitigation costs in developing countries (and the former Soviet Union) on account of their higher emissions intensities, even after output is adjusted for differences in purchasing power. Significant international transfers may be needed to poor, high-mitigation-cost countries. Such compensatory transfers will increase costs for developed countries, and reduce them for developing countries. If the mitigation efforts of developing countries are more cost-sensitive, as seems plausible, this will induce greater total global mitigation.

Combined, these conclusions provide strong backing for mitigation policies that are both equitable (with support for developing countries, which have lower incomes as well as higher costs, and are arguably more mitigation-cost-sensitive) and efficient (to minimize policy costs).

These conclusions also imply that national mitigation costs could lie well above the 1–5 percent of GDP range typically cited. Nevertheless, Dieter Helm's fears with which I began this essay—that the cost of mitigation poses a threat to our standard of living—are misplaced. Even if the actual costs of mitigation were double those in Table 2.1, Figure 2.1 shows that they would still be manageable in the context of a growing economy, at least for most countries. And incurring even heavy costs would be worth it for the avoidance of catastrophic climate change.

We should be worried about the costs of mitigation, not because high costs might make mitigation not worth undertaking, but because high costs, like incomplete participation, will discourage mitigation effort. The costs of mitigation are, one suspects, self-limiting: the higher the average costs, the less mitigation will be undertaken.

Mitigation is a journey into the unknown. This brief survey has revealed that there are large uncertainties, and little historical experience on which to draw. Mitigation might turn out to be cheaper than we think, but there are also serious risks that run the other way. Strenuous efforts to put in place efficient and equitable national mitigation policies will likely have a high environmental return.

ACKNOWLEDGMENTS

This chapter would not have been possible without my involvement in 2008 in the Australian Garnaut Review on Climate Change and, through that, with the Australian Treasury (2008) modelling of the costs of climate change mitigation. I would like to thank Frank Jotzo, Meghan Quinn, and Cath Rowley for comments, as well as participants at the July 2009 conference of the East-West Center and the Korea Development Institute, in particular the discussant for my paper, James A. Roumasset.

NOTES

1. Global cost estimates also typically assume away uncertainty. Jotzo and Pezzey (2007) show that in a stochastic world the use of emissions intensity targets rather than absolute emissions targets can substantially reduce the uncertainty about costs and reduce the net expected costs of global abatement by 15 percent or more relative to the use of absolute emission targets. See McKibbin et al. (2008) on the impact of uncertainty at the national mitigation cost level.
2. That is, 3.9 degrees Celsius above preindustrial levels (Stern 2007:180).
3. Melting of the Arctic summer sea ice and the Greenland ice sheet (with tipping points identified by a survey of experts for both estimated to be below 2° C above the 1990 level); dieback of the Amazon (3–4° C); melting of the west Antarctic ice sheet, disruption of the Atlantic thermohaline circulation, disruption of the Sahara/Sahel and West African monsoon, and dieback of boreal forest (3–5° C); and disruption of the El Niño–Southern Oscillation (3–6° C).
4. Figure 20.3 of IPCC (2007b) reproduces Figure 19.4 from Chapter 19 of IPCC (2001) and adds Stern's estimates, which are within the range of the 2001 estimates.
5. The IPCC Fourth Assessment Report does not go into this issue except to note that with "high prices in the range of 100–150 US$/tCO$_2$ (in 2000 U.S. dollars) more CO$_2$ reductions are expected in China and India than in developed countries when the same level of carbon price is applied" (IPCC 2007a:217).
6. In the model used, there is no significant correlation between a country's marginal mitigation costs (as measured by the extent of the deviation of emissions from business as usual) and its total mitigation costs.
7. "Neither public R&D nor prime reliance on carbon pricing/cap-and-trade will achieve the far-reaching, long-term innovations required to address climate change" (Grubb 2004:120).
8. One estimate of the carbon dioxide price of solar photovoltaics in Germany is 900 euros (http://en.wikipedia.org/wiki/Erneuerbare-Energien-Gesetz#Costs_and_advantages).
9. Certainly, leakage has been the most controversial mitigation policy issue in Australia. In the words of the opposition leader: "That issue of carbon leakage is the problem at the absolute core of this challenge of reducing global carbon dioxide emissions" (Turnbull 2009).
10. The modelling undertaken for the Garnaut Review (2008) was of a mitigation strategy that begins in 2012. Australia's emissions entitlement is reduced in a linear fashion consistent with a 60 percent reduction target by 2050 over 2000 levels. Permits are available (or offsets), but at a premium price due to the absence of comprehensive global mitigation. The MMRF model is used. For more details, see Box 12.2 of the Garnaut Review.

11. Though using a different instrument, McKibbin puts forward a similar argument. Australia should not wait for other countries to act, he argues. Instead a "low short-term permit price can be imposed until other countries are also taking effective action" (McKibbin 2007:11).

REFERENCES

Aldy, J., and W. Pizer. 2009. *The Competitiveness Impacts of Climate Change Mitigation Policies*. Arlington, Virginia: Pew Center on Global Climate Change.

Australian Government. 2008. *Carbon Pollution Reduction Scheme Green Paper*. Canberra: Australian Government.

Australian Treasury. 2008. *Australia's Low Pollution Future: The Economics of Climate Change Mitigation*. Canberra: Australian Government.

Barker, T., M. Qureshi, and J. Köhler. 2006. *The Costs of Greenhouse Gas Mitigation with Induced Technological Change: A Meta-Analysis of Estimates in the Literature*. Working Paper 89. Norwich: Tyndall Centre for Climate Change Research.

Bennear, L.S. and Stavins, R.N. 2007. Second-best Theory and the Use of Multiple Policy Instruments. *Environmental and Resource Economics* 37: 111–29.

Brandt, U., and G. Svendsen. 2006. Climate Change Negotiations and First-Mover Advantages: The Case of the Wind Turbine Industry. *Energy Policy* 34:1175–84.

Brown, Gordon. 2009. Roadmap to Copenhagen. Speech by the Prime Minister of Great Britain, 26 June. www.number10.gov.uk/Page19813, accessed 13 July 2009.

Center for Strategic and International Studies (CSIS). 2007. *The Age of Consequences: The Foreign Policy and National Security Implications of Climate Change*. Washington: Center for Strategic and International Studies.

Consultative Group on International Agricultural Research (CGIAR). 2008. *Bio-fuels Research in the CGIAR: A Perspective from the Science Council*. Washington: CGIAR Science Council Secretariat.

Frankel, J. 2005. "You're Getting Warmer: The Most Feasible Path for Addressing Global Climate Change Does Run through Kyoto." In *Trade and Environment: Theory and Policy in the Context of EU Enlargement and Transition Economies*, edited by J. Maxwell and R. Reuveny, pp. 37–55. Cheltenham, UK and Northampton, MA, USA: Edward Elgar.

Garnaut, R. 2008. *The Garnaut Climate Change Review*. Cambridge and New York: Cambridge University Press.

Grubb, M. 2004. Technology Innovation and Climate Change Policy: An Overview of Issues and Options. *Keio Economic Studies* 41 (2): 103–32.

Hannemann, W. 2007. How California Came to Pass AB 32, the Global Warming Solutions Act of 2006. Paper presented at the conference on Cap and Trade as a Tool for Climate Change Policy: Design and Implementation, University of California, Berkeley, 22–23 February 2007. Department of Agricultural and Resource Economics and the California Climate Change Center, Goldman School of Public Policy, University of California at Berkeley.

Hannemann, W. 2008. A New Architecture for Domestic Climate Policy: Trading, Tax or Technologies? Lecture at the Australian National University, Canberra.

Helm, D. 2008. Climate-Change Policy: Why Has So Little Been Achieved? *Oxford Review of Economic Policy* 24 (2): 211–38.

Hodges, H. 1997. *Falling Prices, Cost of Complying with Environmental Regulations Almost Always Less Than Advertised.* EPI Briefing Paper 69. Washington: Economic Policy Institute.

International Panel on Climate Change (IPCC). 2001. *Climate Change 2007: Contribution of Working Group II to the Fourth Assessment Report of the Intergovernmental Panel on Climate Change.* Geneva: International Panel on Climate Change.

IPCC. 2007a. *Climate Change 2007: Contribution of Working Group III to the Fourth Assessment Report of the Intergovernmental Panel on Climate Change.* Geneva: International Panel on Climate Change.

IPCC. 2007b. *Climate Change 2007: Contribution of Working Group II to the Fourth Assessment Report of the Intergovernmental Panel on Climate Change.* Geneva: International Panel on Climate Change.

Jochem, E., and R. Madlener. 2003. *The Forgotten Benefits of Climate Change Mitigation: Innovation, Technological Leapfrogging, Employment and Sustainable Development.* Paris: Organisation for Economic Co-operation and Development.

Jotzo, F., and J. Pezzey. 2007. Optimal Intensity Targets for Greenhouse Emissions Trading under Uncertainty. *Environmental and Resource Economics* 38 (2): 259–84.

Klaassen, G., A. Miketa, K. Larsen, and T. Sundqvist. 2005. The Impact of R&D on Innovation for Wind Energy in Denmark, Germany and the United Kingdom. *Ecological Economics* 54 (2–3): 227–40.

Kutas, G., C. Lindberg, and R. Steenblik. 2007. *Biofuels: At What Cost? Government Support for Ethanol and Biodiesel in Selected OECD Countries.* Geneva: Global Subsidies Initiative of the International Institute for Sustainable Development. Available in electronic form at www.globalsubsidies.org.

Lenton, T., H. Held, E. Kriegler, J. Hall, W. Lucht, S. Rahmstorf, and H. Schellnhuber. 2008. Tipping Elements in the Earth's Climate System. *Proceedings of the National Academy of Sciences* 105 (6): 1786–93. Available in electronic form at www.pnas.org.

McKibbin, W. 2007. *From National to International Climate Change Policy: The 2006 Sir Leslie Melville Lecture.* Sydney: Lowy Institute for International Policy.

McKibbin, W., and D. Pearce. 2007. *Two Issues in Carbon Pricing: Timing and Competitiveness.* Working Paper in International Economics 1.07. Sydney: Lowy Institute for International Policy. Available in electronic form at www.lowyinstitute.org.

McKibbin, W., A. Morris, and P. Wilcoxen. 2008. *Expecting the Unexpected: Macroeconomic Volatility and Climate Policy.* Brookings Global Economy and Development Working Paper 28. Washington: Brookings Institute.

Nordhaus, W. 2008. *A Question of Balance: Weighing the Options on Global Warming Policies.* New Haven, Connecticut: Yale University Press.

Robb, A. 2009. Parliamentary Speech on the Government's Emissions Trading Legislation, 3 June. Accessed 12 July 2009 at www.andrewrobb.com.au.

Smith, F. 2007. *Corn-Based Ethanol: A Case Study in the Law of Unintended Consequences.* Issue Analysis, 14 June 2007. Washington: Competitive Enterprise Institute.

Smith, Joel B., Stephen H. Schneider, Michael Oppenheimer, Gary W. Yohe, William Hare, Michael D. Mastrandrea, Anand Patwardhan, Ian Burton, Jan Corfee-Morlot, Chris H. D. Magadza, Hans-Martin Füssel, A. Barrie Pittock, Atiq Rahman, Avelino Suarez, and Jean-Pascal van Ypersel. 2009. Assessing Dangerous Climate Change through an Update of the IPCC "Reasons for Concern." *Proceedings of the National Academy of Sciences* 106 (22): 4133–37. Available in electronic form at www.pnas.org.

Stern, N. 2007. *The Economics of Climate Change: The Stern Review*. Cambridge and New York: Cambridge University Press.

Stern, N. 2009. *A Blueprint for a Safer Planet: How to Manage Climate Change and Create a New Era of Progress and Prosperity*. London: Bodley Head.

Toll, R. 2006. The Stern Review of the Economics of Climate Change: A Comment. *Energy and Environment* 17 (6): 977–81.

Turnbull, R. 2009. Speech to the House of Representatives by the Leader of the Opposition, on the Carbon Pollution Reduction Scheme Bill, 2 June. Accessed 13 July 2009 at malcolmturnbull.com.au.

United Kingdom, National Audit Office. 2005. *Department of Trade and Industry: Renewable Industry*. London: National Audit Office.

United Nations. 2009. *World Economic and Social Survey 2009: Promoting Development, Saving the Planet*. New York: United Nations Development Policy and Analysis Division.

Victor, D. 2008. *Climate Accession Deals: New Strategies for Taming Growth of Greenhouse Gases in Developing Countries*. Harvard Project on International Climate Agreements, Discussion Paper 2008-18. Cambridge, Massachusetts: Harvard University.

Weitzman, M. 2007. A Review of *The Stern Review of the Economics of Climate Change*. *Journal of Economic Literature* 45 (3): 703–24.

Weyant, J. 2000. *An Introduction to the Economics of Climate Change Policy*. Arlington, Virginia: Pew Center on Global Climate Change.

3. Tradable carbon allowances: the experience of the European Union and lessons learned

Jos Sijm

INTRODUCTION

In January 2005 the European Union introduced an Emissions Trading Scheme (the EU ETS) in order to reduce greenhouse gas emissions in a cost effective way. Since then, the EU ETS has become the cornerstone of E.U. climate policy, which has attracted much attention and stimulated debate both inside and outside the European Union. The main purpose of the present chapter is to evaluate the performance of the EU ETS during the first three to five years of its existence and to draw some lessons from its experience. These lessons may be useful in particular for other regions or countries interested in setting up and developing their own emissions trading scheme.[1]

The following section outlines some main features of the EU ETS up to 2012. Subsequent sections discuss different aspects of the performance of the EU ETS since early 2005, including the performance of the allocation system, the question of whether the scheme has already led to some carbon abatement, the development of the market for trading E.U. emission allowances, and the impact of the EU ETS on economic growth, industrial competitiveness, and carbon leakage. The next-to-last section discusses some important changes in the fundamentals of the EU ETS, which have been adopted and will be implemented after 2012. The final section provides a summary of some achievements and lessons learned during the first five years of the EU ETS.

MAIN FEATURES OF THE E.U. EMISSIONS TRADING SCHEME UP TO 2012

As part of the Kyoto protocol, the member states of the European Union agreed to reduce their annual greenhouse gas (GHG) emissions over the period 2008–12 by 6 percent on average compared with a reference level of the early 1990s.[2] In order to meet this commitment—and even more ambitious GHG mitigation targets beyond 2012—the European Union decided to establish the EU ETS, which started to operate from January 2005. The main features of this scheme up to 2012 are outlined below.

Type of system and operational rules
The EU ETS is a so-called cap-and-trade system including the following characteristics and operating rules.

- An absolute quantity limit (or cap) is set on the total CO_2 emissions of the participating installations over a certain period. This limit or quantity of emission *allowances* is allocated among these participants, which are allowed to trade these allowances among each other.[3]
- Participants have to monitor and annually report their carbon emissions, verified by external parties.
- On 30 April each year, participants have to surrender a quantity of allowances equal to their verified emissions in the preceding calendar year.
- Noncomplying participants have to pay a penalty for each metric ton of CO_2 not covered by surrendered allowances. This penalty amounts to 40 euros per ton of CO_2 (tCO_2) during the first phase of the EU ETS (2005–07) and 100 euros per tCO_2 during the second phase (2008–12). In addition, the names of these noncompliers are listed ("naming-and-shaming"), and they have to surrender allowances in the next year for these noncompliance emissions.

Timing, trading, banking, and borrowing
Up to 2012 the EU ETS is distinguished by two trading periods. In phase 1 (2005–07), the primary purpose of this pilot or trial period was to develop the EU ETS infrastructure and to gain experience to improve the system in subsequent periods. Phase 2 (2008–12) corresponds to the commitment period of the Kyoto protocol.

Within any trading period, there is effectively no restriction on trading, banking, or borrowing of allowances. Although allowances are issued annually, they are valid for covering emissions in any year within the trading period. Between the first and the second trading period, however, neither effective banking nor borrowing was possible. For the second and subsequent trading periods, on the other hand, unrestricted *inter*-period banking—but no borrowing—is allowed (European Commission 2003; Ellerman and Joskow 2008).

Coverage
During the first phase, the EU ETS covered only CO_2 emissions from specified sectors and activities, including (1) all large installations in the electric
power and heat sector with a thermal input of greater than 20 megawatts
(MW) as well as all other combustion plants (more than 20 MW input),
regardless of the sector in which they are found, including commercial and
institutional establishments, and (2) specified installations, meeting certain
input or output capacity thresholds, in selected energy-intensive industries,
including oil refineries, coke ovens and plants making iron and steel, cement,
lime, glass, bricks, ceramics, paper, and pulp.[4] Overall, the EU ETS covers
about 11,000 installations which are collectively responsible for approximately 2 gigatons (Gt) of CO_2 emissions, that is, nearly half of the European
Union's CO_2 emissions and some 40 percent of its total GHG emissions.

Allowance allocation
Up to 2012, both cap-setting and the distribution of allowances at the
installation level are the primary responsibility of individual member
states, which have to design national allocation plans (NAPs) for each
trading period based on criteria and guidelines set by the European
Commission. Allowances should be allocated free of charge, but member
states have the option to auction up to 5 percent of allowances in phase
1 and up to 10 percent in phase 2. Based on the E.U.-wide allocation
criteria and guidelines, the NAPs have to be judged and approved by the
European Commission, which has the option to suggest NAP adjustments
to member states before giving its final approval.

External linkages
In order to meet their obligations under the EU ETS, participants may
convert a limited amount of offset credits from joint implementation and
clean development mechanism (JI and CDM) projects into additional
E.U. allowances. This amount is restricted to a certain percentage of the
allocated allowances, which may vary between member states and sectors.
All types of JI and CDM credits are allowed for conversion, except credits
from nuclear facilities and carbon sinks. In addition to links with JI and
CDM markets, the E.U. directive on emissions trading offers the opportunity to set linkages between the EU ETS and other (future) schemes across
the world in order to develop and stimulate a more cost effective, global
carbon market.

Allowances transaction registries
Allowances are not printed but held in electronic accounts in national
registries set up by member states. Through legislation, the European

Commission has set up a standardized and secured system of registries based on U.N. data exchange standards to track the issuance, holding, transfer, and cancellation of allowances. Provisions on the tracking of JI and CDM credits in the E.U. system are also included (European Commission 2007).

At the E.U. level, the system of national registries is overseen by a central administrator and connected to a central registry in Brussels: the Community Independent Transaction Log (CITL). In addition to recording transactions of E.U. allowances (EUAs) among installations in different member states, the CITL also provides data on free allocations and verified emissions at the installation level as reported by the member states. Finally, the E.U. registries system is connected to and will be further integrated with the international registries system under the United Nations Framework Convention on Climate Change (UNFCCC).

THE PERFORMANCE OF THE ALLOCATION SYSTEM

This section evaluates the allocation system of the EU ETS up to 2012. To provide a frame of reference for our evaluation, we start with a brief discussion of the standard ("ideal") allocation system from the textbook literature. Subsequently, we outline some actual characteristics of the EU ETS allocation system during its first and second trading periods, indicate some of its differences (compared with the standard system), and finally discuss some implications of these characteristics and differences for the performance of the EU ETS allocation system.

The standard allocation system

In the standard cap-and-trade system, a central authority imposes a cap on total emissions and then allocates the total number of emission allowances to the participants, where the cap equals the number of allowances. Basically, two reference cases of allocating emission allowances can be distinguished, that is, auctioning and perfect free allocation. In an auctioning system, allowances are initially allocated by selling them at an auction (or market). On the other hand, the ideal (textbook) type of perfect free allocation is characterized by the following. (1) There is a one-off initial allocation of free allowances to existing installations (incumbents), usually for a long time frame, based on (a) a fixed baseline or historic reference period of actual emissions at the installation level ("grandfathering") or (b) a standard emission factor multiplied by an ex-ante fixed quantity or activity level, for instance a certain input, output, or capacity level ("benchmarking"). (2) At closure, installations retain their allowances. (3) New entrants do not receive allowances for free, but have to buy them on the market.

The resulting market equilibrium price of an allowance will determine the actual pattern of emissions and abatements among the sources covered by the scheme. A cost-minimizing source will purchase (or sell) allowances until its marginal abatement costs (MAC) equal the market price of an allowance. This implies that, in equilibrium, the MAC of all sources will be equal, which satisfies the necessary condition for minimizing the total abatement costs of realizing the cap (Kruger et al. 2007).

Moreover, as the initial distribution of emission allowances in a perfect free allocation system is independent of a plant's operation, closure, and investment decisions, it creates the same set of conditions for abatement efficiency as an auctioning system (Harrison et al. 2007). Hence, both allocation systems result in the same level of the allowance price, the same level and type of abatement, the same marginal and total abatement costs, and the same level of passing-through costs to output prices. The only difference between auctioning and perfect free allocation concerns the transfer of economic rent due to the initial allocation of emission allowances. Whereas this wealth accrues to the central authority in the case of auctioning, it is transferred to the recipients of allowances in the case of perfect free allocation (Neuhoff, Keats, et al. 2006).

Characteristics of E.U. allocation up to 2012
During the first and second trading periods, the allocation system of the EU ETS showed some differences compared with the standard outline above, in particular (1) the decentralized structure of E.U. allocation decisionmaking, (2) the incidence of some specific free allocation provisions, and (3) the relatively short duration of the allocation periods. More specifically, the major characteristics of the E.U. allocation system up to 2012 include the following.[5]

A decentralized structure of EU ETS allocation decisionmaking Up to 2012, some key parts of E.U. decisionmaking on allocation issues are rather decentralized, as both national cap-setting and allocation of allowances to installations are the primary responsibility of individual member states, which have to address these issues in NAPs for each trading period. These NAPs, however, have to be judged and approved by the European Commission, notably to determine whether they meet certain E.U.-wide allocation criteria and guidelines (which to some extent restrict the room for national decisionmaking).[6] This more decentralized structure of the E.U. allocation system is due to some factors inherent to the European

Union, including (1) the still high degree of autonomy of E.U. member states to decide on economic and environmental issues, (2) the lack of a commonly acceptable methodology and database to allocate allowances to installations in a harmonized way across the European Union, and above all (3) the large diversity and heterogeneity among E.U. member states, notably the major differences in efforts needed to meet their Kyoto targets (including differences in sharing this target between trading and nontrading sectors as well as differences in relying on imports of JI and CDM credits).

Relatively short allocation periods　The allocation periods of the EU ETS, which corresponded to its trading periods, are relatively short: only three years during the trial phase (2005–07) and five years during the second phase (2008–12).

Free allocation based on grandfathering and projections　Although member states had the option to auction up to 5 percent of their allowances in phase 1 and up to 10 percent in phase 2, only a few countries auctioned even a tiny share of their allowances.[7] Almost all member states allocated up to 100 percent of their allowances for free. For existing installations (incumbents), free allocation was usually based on recent historical emissions and projections of growth rates of business-as-usual emissions (in order to allocate as many allowances as needed to internationally competing, non-power sectors). Expected shortages of allowances were usually allocated to the power sector, notably during phase 2, based on the assumption that this industry, compared with other trading sectors, generally has cheaper abatement options and is highly protected from outside competition and hence is better able to pass on its ETS costs to output prices.

Specific free allocation provisions for new entrants and plant closures　All member states have set up reserves for allocating free allowances to new entrants, and most require closed facilities to forfeit post-closure allowances allocated for free. In contrast to the reference standard (or other, comparable schemes elsewhere), these provisions are highly novel. They have been adopted in order to prevent disfavoring the European Union in competition for new investments and to eliminate an incentive to shut down facilities and move production elsewhere (Convery et al. 2008; Sijm et al. 2008a). Among the member states concerned, however, the specific rules of the free allocation provisions—notably for new entrants—varied widely. This resulted in significant differences in the amount of free allowances allocated to similar new investments across E.U. countries (Åhman and Holmgren 2006).

Specific problems   During the allocation process for the first trading period, member states faced some specific problems, including (1) tight time schedules in preparing their NAPs, (2) unclear definitions of the types of installations to be covered by the scheme, (3) lack of uniform, consistent, and reliable emissions projections, and above all (4) lack of reliable installation-specific emissions data (Convery et al. 2008). For the second phase, these problems were largely overcome, mainly owing to (1) slightly more relaxed time schedules, (2) more consistent guidance by the Commission, including clear definitions of coverage, (3) the availability and use of a single, transparent, and consistent model to project growth of emissions for all member states, and (4) the availability of 2005 verified emissions data at the installation level. Nevertheless, also for the second period, the whole cycle from first preparations to final approvals of 27 NAPs was a cumbersome process that was highly resource- and time-demanding, including the consultations between the member states and the national stakeholders on the one hand and between the member states and the European Commission officials on the other hand.

Implications of decentralized allocation up to 2012
The characteristics and underlying conditions of the E.U. allocation system up to 2012, as outlined above, had some major implications for the performance of the EU ETS. The major advantage of the rather decentralized structure of allocating EUAs was that it could meet relevant differences in socioeconomic conditions among member states, notably major differences in efforts needed to meet differentiated Kyoto targets. On the other hand, this structure also had some adverse implications, as discussed below.

Inefficient allocation of abatement targets between trading and nontrading sectors   The decentralized structure of E.U. allowance allocation implies that each member state individually determines what share of its national emissions budget it will allocate to its ETS sectors. Thus, each country is effectively creating a certain number of EUAs and the aggregate supply of EUAs (that is, the E.U.-wide cap) is the sum of these allocations for all the member states. This structure of decentralized EUA supply decisions, however, implies that for any member state it is hard to predict the EUA market price at the time that it sets its own NAP, since it would have to know all the other NAPs in advance. Hence, it is difficult for any member state to set the most efficient allocation of its national emissions budget between the trading and nontrading sectors (Kruger et al. 2007).

Each member state, on the other hand, will be inclined to protect its internationally competing (ETS) sectors and, hence, allocate free allowances amply to these sectors (in particular as it is uncertain about what competing member states will do but, most likely, will also treat these sectors favorably). Therefore, the decentralized structure of the EUA allocation system is likely inclined to result in an over-allocation of allowances to the ETS sectors and hence in less efficiency in overall abatement (notably as abatement options in the trading sectors are generally assumed to be cheaper than in the nontrading sectors).

Over-allocation of E.U. allowances to the (internationally competing) ETS sectors seems indeed to have happened in most member states particularly during the first phase of the scheme, even after the Commission had reduced the amount of CO_2 allowances in 15 NAPs by 290 million tons annually in total (Clo 2009; see also the section on abatement or over-allocation below). Apart from the decentralized character of the EUA allocation process, however, this over-allocation was due also to the fact that the first trading period was a trial phase in which the abatement target of the scheme was not ambitious and, at least for this period, there was no Kyoto or other national mitigation target for the member states.

However, also for the second trading period (in which the Kyoto targets applied), most of the draft NAPs proposed by member states showed indications of major over-allocations of EUAs to ETS sectors compared with non-ETS sectors, based on certain proportional standards between these two groups of sectors (Clo 2009). Only after the Commission had reduced the amount of EUAs in 23 NAPs-II by more than 240 million tons annually in total, did there seem to be a reasonably proportional sharing of the overall abatement burden between the trading and the nontrading sectors.[8]

Race-to-the-bottom effect   Due to the decentralized structure of the EUA allocation process, each member state was largely uncertain about the allocation decisions of other member states. Hence, each was inclined to take the safe side of its own decisions based on national considerations rather than on the optimal outcome for all member states as a whole (that is, the so-called "prisoner's dilemma" or "race-to-the-bottom" effect). This applies not only to decisions to allocate national emissions budgets favorably to (internationally competing) ETS sectors, as discussed above, but also to decisions to allocate up to 100 percent of the allowances free of charge (and not or hardly to auction at all) as well as to the widely accepted free allocation provisions for new entrants and plant closures.

Equity and competitive distortions   On the other hand, the decentralized EUA allocation structure and the large differences among member states in socioeconomic conditions, particularly in meeting their Kyoto targets, led to significant differences in allocation to similar installations in different countries. In turn, this resulted in (widespread complaints on) competitive distortions among these installations—notably with regard to decisions on plant closures and new investments—as well as in uneven equity and liquidity effects among existing installations.

Implications of free allocation up to 2012
The most important advantage of the provision to allocate at least 90–95 percent of the allowances up to 2012 free of charge was that it facilitated the introduction of the EU ETS, as it made the scheme more acceptable to both member states and stakeholders. On the other hand, free allocation resulted also in some contentious or adverse effects.

Windfall profits Free allocation led to the (putative) incidence of so-called "windfall profits" due to either (1) the over-allocation of free allowances to industrial installations (which they could sell on the market) or more particularly (2) the pass-through of the opportunity costs of free allowances, notably in the power sector, resulting in higher electricity prices and generators' profits (Sijm et al. 2006; Sijm et al. 2008b). This incidence of windfall profits, which led to sometimes fierce controversies, undermined the widespread acceptability and credibility of the EU ETS, notably as some questioned also the environmental effectiveness of the scheme. (See also the section on abatement or over-allocation below.)

Perverse incentives Emissions trading with free allocation provisions for new entrants and/or plant closures can be regarded as a subsidy toward the investors' fixed costs, coupled with an emissions tax on their variable costs. While the tax encourages cleaner production, the subsidy gives an incentive to invest in additional dirty capacity, or to refrain from closing existing, more polluting capacity, or both. Moreover, as these provisions during the first and second phases of the EU ETS were usually fuel-specific (that is, dirty installations get more free allowances), they actually provide a perverse incentive for higher emissions, thereby undermining the carbon efficiency and environmental integrity of the scheme (Sijm et al. 2008a).

Rent-seeking Free allocation encourages all kinds of lobbying, gaming, and other rent-seeking activities—including promoting demands for all kinds of special allocation rules and exemptions for particular groups—thereby further enhancing the diversity, complexity, and lack of transparency of national allocation plans.

Implications of short allocation periods up to 2012
One of the major disadvantages of the relatively short allocation periods of the EU ETS up to 2012 (that is, 3–5 years) is that it offers little certainty on allocation issues beyond these periods to investments in the power and energy-intensive industries, which often have a lifetime of 30 to 50 years or more. Another disadvantage is that it enhances both the need and risk of frequently updating the baseline period for allocating free allowances to existing installations. This provides an incentive to these incumbents to inflate their present emissions in order to receive more free allowances in the future and, hence, reduces the carbon efficiency of the scheme.

On the other hand, while phase 1 allocation was characterized by a number of problems, it is important to note that the relatively short first phase of the EU ETS was above all a trial period aimed at gaining lessons, insights, and data that could be used to improve allocation during subsequent periods. Indeed, some findings and lessons learned during the first phase were used to improve allocation in the second phase. In particular, besides using the verified 2005 emissions database, the Commission harmonized certain allocation rules, strengthened certain allocation guidelines and tightened the carbon constraint in phase 2 (Convery et al. 2008). More importantly, the lessons from the review of phase 1 were used to drastically revise the EU ETS directive for the third period and beyond, in particular to substantially improve the performance of allowance allocation after 2012 (see the section on changes beyond 2012 below).

ABATEMENT OR OVER-ALLOCATION?

The ultimate aim of the EU ETS is to reduce emissions. This raises the question of whether the scheme has already resulted in some carbon abatement during its trial phase, or whether this period was too short and the allocation of allowances too generous—leading to low and falling carbon prices down to zero by the end of this phase—and hence induced hardly any CO_2 reduction. This question of "abatement or over-allocation" is addressed in the present section. The first of the subsections below deals briefly with the definition of the concept of over-allocation, and the second summarizes some findings on the balance of allocated allowances and verified emissions at the installation-, sector-, country-, and E.U.-wide level during the first two years of the scheme (2005–06). The next subsection reviews some recent studies on allocation and abatement in the EU ETS during phase 1, and the section ends with a discussion and some conclusions.

Defining over-allocation

As noted by Ellerman and Buchner (2008), over-allocation is usually not a well-defined concept. It refers to the notion that too many allowances were allocated, but the standard by which "too many" is to be determined is rarely specified. They suggest two standards of reference. The first is what emissions would have been without the trading scheme: the so-called "counterfactual" or what is termed "business as usual" (BAU) emissions in modeling exercises. According to this standard, issuing more allowances than BAU emissions would constitute over-allocation.

The second standard refers to a cap that is constraining: in other words, less than the counterfactual, but still judged not sufficiently ambitious. For instance, if the desired degree of ambition were a 5 percent reduction of emissions from the counterfactual, and allowances were issued such as to require only 2 percent reduction, the 3 percent difference might be considered over-allocation (Ellerman and Buchner 2008). Although sometimes poorly specified, this second definition is often used in much of the current debate, while the first definition seems to be more common in the recent academic literature on analyzing and estimating over-allocation in the EU ETS (see the subsection on estimates of phase 1 below).

In both definitions, however, over-allocation is hard to estimate, as both involve the construction of a counterfactual estimate of what emissions would have been in the absence of the EU ETS. This counterfactual estimate should take into account variables such as economic growth, energy prices, weather conditions, and non-ETS policies, since all of these variables affect what emissions would have been without emissions trading (Ellerman and Joskow 2008). Similar difficulties apply to estimating the level of abatement due to the EU ETS, as such an exercise also involves a counterfactual estimate of emissions in the absence of the EU ETS. Nevertheless, despite these difficulties, some recent studies have tried to analyze and estimate the level of over-allocation and/or abatement in the first phase of the EU ETS (see subsection below).[9]

Allocated allowances and verified emissions: 2005–06 results

In a detailed study, Kettner et al. (2008) have analyzed the net positions between allocated allowances and verified emissions of almost 10,000 installations in the EU ETS for 2005 and 2006, based on data from the CITL. Their major findings include the following.[10]

- In the two trading years, the EU ETS as a whole was in a net long position: the number of allocated allowances exceeded verified emissions by some 70 million tons of carbon dioxide ($mtCO_2$) per year on average (about 3.4 percent of the total allocations in these years).[11]

- Of the 9,900 installations reported up to May 2007 in the CITL database, almost 2,700 were short. The net positions of installations, however, varied between member states and sectors.

- Of the 24 member states analyzed, only 5 countries were short in 2005–06, ranging from Austria (−1.1 percent) to the United Kingdom (−17.4 percent), with intermediate positions for Italy (−7 percent), Spain (−7.6 percent), and Ireland (−15.6 percent). The remaining countries were long (up to almost 46 percent in Lithuania). In absolute amounts, the countries with the largest net long positions in 2005–06 were Poland (on average, a surplus of 31 $mtCO_2$ per year), Germany (25 $mtCO_2$), France (20 $mtCO_2$), and the Czech Republic (14 $mtCO_2$).[12] Together, these four surplus countries supplied the bulk of the net demand from the five deficit countries, implying significant EUA transfers versus net capital flows between the installations of these countries.

- At the E.U.-wide sector level, only power and heat was short, with a net position in 2005–06 amounting to 44 $mtCO_2$ per year on average (that is, 4 percent of the allowances allocated to this sector). All the other sectors recorded significant net long positions, notably pulp and paper (19.6 percent), iron and steel (17.5 percent), and ceramics (17.3 percent).

A related, interesting finding by Kettner et al. (2008) refers to the pronounced inequality of the distribution of the size of installations when ordered according to their verified emissions in 2005–06. The smallest three-quarters of all installations contribute only about 5 percent of all emissions covered by the EU ETS, whereas the biggest 1.8 percent of all installations account for half of the emissions. The 1,000 biggest installations, or one-tenth of all installations, are responsible for 86 percent of the EU ETS emissions.

Estimates of phase 1 over-allocation and abatement

The findings on the net EUA positions of installations during the first phase of the EU ETS as well as the resulting fall of the price of phase 1 EUAs toward zero in 2007 (see the next section below) have fuelled controversies on whether the system was actually "over-allocated" and, in addition, whether and to what extent it has contributed to carbon abatement during this phase. Some recent studies have tried to analyze the level of over-allocation and/or abatement in the first period of the EU ETS. The major findings of these studies are discussed briefly below.

Ellerman and Buchner   Ellerman and Buchner (2008) were the first to analyze whether the 2005–06 emissions data of the EU ETS reveal over-allocation or abatement. They conclude that both occurred in each year. More specifically, they note that 2005 and 2006 emissions were lower than the historical baseline emissions used in the development of the first NAPs, despite continuing economic growth in the European Union and increases in oil and natural gas prices that could be expected to increase the demand for coal-fired power generation. Using a simple counterfactual based on the extrapolation of trends in pre-2005 emissions, economic growth, energy use, and CO_2 intensity, they conclude that abatement in 2005–06 was probably between 50 and 100 $mtCO_2$ in each of these years (that is, between 2 and 5 percent of covered emissions). In addition, they find that over-allocation occurred and that its magnitude may have been as much as 6 percent or 125 million EUAs per year (Ellerman and Buchner 2008; see also Ellerman and Joskow 2008).

Delarue and others   Delarue, Ellerman, and D'haeseleer (2008) and Delarue, Voorspoels, and D'haeseleer (2008) use a simulation model of the E.U. power sector ("E-simulate") in order to estimate short-term abatement through fuel switching in this sector in response to the introduction of a CO_2 allowance price during the first phase of the EU ETS. Their estimates of the lower and upper bounds of this type of abatement vary between 34 and 88 $mtCO_2$ in 2005 and between 19 and 59 $mtCO_2$ in 2006. Abatement through fuel switching is shown to depend not only on the EUA price but also, and more importantly, on the load level of the system, the ratio between the natural gas and coal prices, and the availability of natural gas generating capacity. They show that most of the estimated abatement due to the EU ETS in 2005 and 2006 occurred in the United Kingdom and Germany, where a significant reliance on coal is coupled with available natural gas generating capacity.

Ellerman and Feilhauer   The study by Ellerman and Feilhauer (2008) uses top-down trend analysis and a bottom-up sector model to define upper and lower boundaries on abatement in Germany in the first phase of the EU ETS. Differing emission intensity trends and emission counterfactuals are constructed using emissions, power generation, and macro economic data. Resulting top-down estimates set the upper bound of abatement in phase 1 at 122 $mtCO_2$ for all ETS sectors and 57 $mtCO_2$ for the power sector only. Using a tuned version of the model "E-simulate" (similar to the model applied by Delarue et al., mentioned above), a lower boundary of phase 1 abatement is established at 13 $mtCO_2$, based only on fuel switching in the power sector (which constitutes 61 percent of German ETS sector emissions).[13]

Widerberg and Wrake   Widerberg and Wrake (2009) analyze the short-term impact of the EUA price on CO_2 emissions from power generation in Sweden, using an econometric time series analysis for the period 2004–08. They control for effects of other input prices and hydropower reservoir levels. Their results do not indicate any link between the EUA price and the CO_2 intensity of Swedish electricity production. This result may be explained by a number of reasons, in particular the fuel mix capacity and other structural characteristics of Swedish power generation. Overall, they draw two main conclusions.

> First, it seems unlikely that the EU ETS has generated any significant reductions of CO_2 emissions in Swedish electricity generation. Second, it seems unlikely that there are significant volumes of low-cost CO_2 abatement measures with short response times in the Swedish electricity sector. In order to better understand the long-term impacts of the EU ETS on CO_2 intensity, one needs to complement the analysis with studies that have stronger emphasis on investment planning.

Anderson and di Maria   Anderson and di Maria (2009) used dynamic panel data techniques to assess the level of abatement and over-allocation that took place across European countries during the pilot phase of the EU ETS. In addition to gross over-allocations of 469 $mtCO_2$ for the period 2005–07 as a whole, they also found under-allocations amounting to 211 $mtCO_2$, resulting in net over-allocations of 258 $mtCO_2$. On the other hand, they estimated total abatement during the trial period at 117 $mtCO_2$ as a whole: about 40 $mtCO_2$ per year or approximately 2 percent of the overall cap. However, due to the allocation methodology of the member states and possible uncertainty about future allocation, Anderson and di Maria also found a so-called "emissions inflation" estimated at 230 $mtCO_2$ for the years 2005–07.[14] Hence, they conclude that, on balance, emissions during the trial period of the EU ETS were approximately 113 $mtCO_2$ higher than they would have been in the absence of the EU ETS.

Over-allocation and abatement: discussion and conclusion
The findings of some recent studies on allocated allowances and verified emissions during the first phase of the EU ETS seem to indicate that the scheme most likely resulted in both over-allocation and abatement, up to a few percent of total emissions covered. It is important to keep in mind, however, that this phase was above all a trial period in which the cap was set at a moderate target. Moreover, this phase was characterized by tight

time schedules for designing NAPs and lack of reliable data and projections on emissions at the installation or sector level. In that sense, it is not strange that the first phase of the scheme resulted in over-allocation in a number of sectors and countries, particularly in the internationally competing, industrial sectors and in some East European countries.

The incidence of over-allocation, however, is likely to be reduced substantially—or even eliminated completely—during the second or subsequent trading periods, due to improved emissions data, a more stringent cap and, on average, a continuing growth of economic activities and related emissions.[15]

Moreover, as noted above, despite signs of over-allocation and a moderate target, there are also indications and study findings showing some carbon abatement already during the first phase of the EU ETS. These findings are supported by the observations that (1) the EUA price was, on average relatively high during 2005 and 2006 (15–20 euros per tCO_2, providing a major incentive for carbon abatement) and (2) 2005 and 2006 emissions were lower than the historical baseline emissions used in the design of the first NAPs, despite continuing economic growth in the European Union and increases in oil and natural gas prices that could be expected to increase demand for coal-fired power generation (Ellerman and Buchner 2008).

Besides some small energy efficiency improvements across the scheme, abatement during the first phase was most likely restricted to short-term fuel switching in the power sector of some countries (that is, Germany and the United Kingdom), including switching from lignite to coal or from coal to either gas or biomass. In the medium and long term, however, when the cap becomes more binding and the EUA price higher, the EU ETS will most likely induce other types of abatement, including new investments in carbon saving technologies (and further R&D) across all ETS sectors and reduced demand for electricity and other carbon-intensive goods, due to the pass-through of carbon costs to output prices. Future empirical studies have to reveal whether and to what extent the EU ETS has indeed resulted in these other, more significant types of abatement.

THE DEVELOPMENT OF THE E.U. ALLOWANCE MARKET

Market infrastructure and transactions
Bilateral forward trades in EUAs had already begun in the spring of 2003, well before the official start of the scheme in January 2005. During the next five years, the market for trading EUAs developed strongly in terms

of market infrastructure and transactions. The spot market was launched in early 2005 when the first national registries entered into operation.[16] Trading in standardized contracts for spot or forward markets started in mid-2005, when the first organized marketplaces were set up, followed by trades in futures such as swaps or options in subsequent years. Unlike registries, the development of these marketplaces was the result of voluntary, private initiatives undertaken primarily by energy market managers (Convery et al. 2008). Six marketplaces were launched in 2005: the European Climate Exchange (EXX), Nordpool, Powernext (now Bluenext), the European Energy Exchange (EEX), the Energy Exchange Austria (EEA), and Climex. Besides offering standardized contracts for spot or future delivery with public bids and asks, they also provide clearing services that may be used in confidential over-the-counter (OTC) transactions: bilateral transactions between participants or transactions via banks or brokers such as Natsource, Evolution Markets, or CO2e. com.

Table 3.1 shows that over the years 2005–08 the EUA market grew rapidly and that it dominated the global carbon market by far. In terms of volume of transactions, it increased almost tenfold from 320 mtCO$_2$eq in 2005 to 3,100 mtCO$_2$eq in 2008. In terms of value of transactions, the expansion of the EUA market has been even more impressive over these years: from almost US\$8 billion to US\$92 billion. As a share of global carbon market transactions in 2005–08, the EU ETS accounted for some 45–69 percent in volume terms and for 75–78 percent in value terms.

The EU ETS, however, has been responsible not only for the rapid development of the EUA market but also for the promotion of the JI and CDM market through its provision to meet system compliance by means of JI and CDM credits. Table 3.1 shows that, besides EUA trades, JI and CDM transactions in 2005–08 accounted for a major share of the global carbon market in these years. A major part of these transactions is due to forward purchases of JI and CDM credits by EU ETS installations for either phase 2 compliance or other, risk-hedging, and financial purposes.

Evolution of the E.U. allowance price
Figure 3.1 shows the evolution of the EUA price on the forward market over the period from July 2004 to July 2009. As no banking or borrowing of allowances was allowed between the first and second trading periods, a distinction is made between the forward EUA price for phase 1 allowances (with delivery in December 2006/2007) and phase 2 allowances (with delivery in December 2008/2009).

In addition to institutional factors (such as banking or borrowing rules)

Table 3.1   Carbon market at a glance: volumes and values, 2005–08

Item	Volume (million tCO$_2$eq)				Value (US$ million)			
	2005	2006	2007	2008	2005	2006	2007	2008
Allowance markets								
EU ETS	321	1104	2060	3093	7908	24436	49065	91910
New South Wales	6	20	25	31	59	225	224	183
Chicago Climate Exchange	1	10	23	69	3	38	72	309
Regional Greenhouse Gas Initiative	na	na	na	65	na	na	na	246
Assigned amount units	na	na	na	18	na	na	na	211
Subtotal	328	1134	2108	3276	7971	24699	63007	92859
Project-based transactions								
Primary CDM	341	537	552	389	2417	5804	7433	6519
Joint implementation	11	16	41	20	68	141	499	294
Voluntary market	20	33	43	54	187	146	363	397
Subtotal	372	586	636	463	2673	6091	8195	7210
Secondary CDM								
Subtotal	10	25	240	1072	221	5804	5451	26277
Total	710	1745	2984	4811	10562	31235	63007	126345
EU ETS share of total (%)	45	63	69	64	75	78	78	73

Note: Data do not necessarily add up to the respective column totals, because of rounding errors.

Sources: World Bank (2007, 2008, and 2009).

and market imperfections (for example, lack of information or the use of market power), the EUA price is governed basically by the balance of EUA supply and demand. As no JI and CDM credits could be traded during the first period and installations could not borrow allowances from the second period, EUA supply during phase 1 was simply equal to the EU ETS cap of allocated allowances for the years 2005–07. Due to the inability to bank phase 1 allowances for the second period, EUA demand was driven by actual and anticipated emissions during phase 1 of the scheme, which depended on economic growth, weather conditions, relative energy prices, non-ETS policies affecting ETS emissions and marginal abatement costs, and potentials of carbon reduction options. These different drivers

euros per ton of CO_2

day, month, and year

Source: Nordpool ASA database.

*Figure 3.1   Evolution of the carbon price of an E.U. allowance, July 2004
to September 2009*

can largely explain EAU pricing during the first trading period, which was
marked by three stages (Convery et al. 2008):[17]

The launch period (January 2005–April 2006)   During this stage, the
power sector immediately started buying the EUAs it needed, whereas
many industrial players with surplus allowances were not able or pre-
pared to sell their EUAs. Demand from power producers rose over the
period owing to increased gas prices during the winter. This created scarcity and increased EUA prices. The information available on the market

was very poor, and most of the participants expected an overall short market.

The information shock (April–May 2006) In April 2006, the European Commission released the 2005 verified emissions data for the installations covered by the EU ETS, which showed a 4 percent surplus of allowances. This information hit EUA prices hard, as the supposed scarcity of allowances confronted the reality of a surplus.

Total disconnection between phases 1 and 2 allowance prices (since November 2006) EUA prices for phase 1 started to converge toward zero, reflecting the surplus of allowances over 2005–07 and the inability to bank EUAs for subsequent periods. On the other hand, EUA prices for phase 2 remained relatively steady and rose to as much as 25 euros per tCO_2 in response to the European Commission's stricter review of second period NAPs and the European Council's decision to reduce E.U. emissions to 20 percent below 1990 levels by 2020 (compared with 8 percent below 1990 levels in 2008–12).

Major achievements and lessons
Since 2005, the EUA market has developed strongly in terms of market structure and transactions. Probably its most important achievement during phase 1 has been that it made stakeholders realize that carbon emissions have a price to be included in their decisionmaking. In addition the experience of this phase provides at least the following two lessons: (1) market efficiency and price stability depend on market participants' ability to access timely and reliable information and (2) the decision not to allow inter-period banking strongly contributed to the price volatility of phase 1 allowances and resulted in a full disconnection of EUA prices between the first two periods of the scheme (Convery et al. 2008).

IMPACT OF THE EU ETS ON ECONOMIC GROWTH, INDUSTRIAL COMPETITIVENESS, AND CARBON LEAKAGE

In a specific region or country, emissions trading may have a significant impact on economic growth in general and industrial competitiveness in particular, notably if similar policies are not implemented in other regions or countries. Although up to now the possible impact of the EU ETS on economic growth has received hardly any attention, a major part of the literature and stakeholders' discussions has focussed on the potential impact

Table 3.2   Average GDP growth in OECD countries, 2001–04 versus 2005–08

Country or region	Average 2001–04	Average 2005–08
European Union (EU-27)	1.8	2.2
Japan	1.2	1.4
Korea	4.6	4.1
Mexico	1.5	3.3
United States	2.1	2.2
Total OECD	2.0	2.3

Sources: Eurostat (2009) and OECD (2009).

of the scheme on industrial competitiveness and the related concept of "carbon leakage." These issues will be addressed in the present section.

The first subsection below analyzes briefly whether some impact of the EU ETS on the GDP growth performance of the EU-27 can already be observed. The second subsection discusses the concepts of industrial competitiveness and carbon leakage. The third subsection tries to identify sectors at risk of carbon leakage and loss of competitiveness due to unilateral climate policies. The final subsection evaluates very briefly the evidence on carbon leakage in industrial sectors due to the EU ETS.

Economic growth in E.U. member states

Table 3.2 provides a summary of the average GDP growth performance of the EU-27 and some other OECD countries over the period 2001–08, distinguished by the four-year period 2001–04 prior to the EU ETS and the four-year EU ETS period 2005–08. For the EU-27, this performance amounted to an average annual growth rate of 1.8 percent in 2001–04 and 2.2 percent in 2005. In comparison, for the United States—which up to now has implemented hardly any similar climate policy—these figures were 2.1 and 2.2 percent, respectively. Hence, at first sight, these data do not indicate a significant impact of the EU ETS on the growth performance of the EU-27.

Some qualifications, however, can be added to this finding. First, the EU-27 growth rate for the period 2005–08 does not show what this rate might have been in the absence of the EU ETS. Although constructing such a counterfactual is quite complicated, it may reveal that the EU ETS had some (negative) impact on economic growth in the EU-27 over the years 2005–08.

Second, the two four-year periods considered are relatively short and, hence, the average growth rate for these periods may result from incidental or accidental factors, such as one or two years with relatively high (or low) growth rates related to global economic conditions. Rigorous future studies, covering and comparing much longer time periods, may provide a better insight into the possible impact of the ETS on the growth perform-ance of the EU-27 (or on the differentiation of this performance among individual member states).

Third, during the years 2005–08 the reduction target of the EU ETS was still modest. Hence, the (negative) impact of this scheme may become more significant in the long run when the cap becomes more binding. However, according to model estimates for the impact assessment of the ambitious E.U. energy and climate policy package proposed in 2008 for the period up to 2020 (which includes a major revision and strengthening of the EU ETS beyond 2012, as discussed in the next section below), the overall effect of this package would be a small reduction of GDP in the EU-27 by only 0.2–0.5 percent in 2020 (Delbeke et al. 2009). This result may be due to the fact that, besides negative growth effects, the policy package may also have positive effects due to induced improvements in energy efficiency or the promotion of renewables and other carbon saving technologies.

Nevertheless, regardless of these qualifications, the average growth rate for the EU-27 in the years 2005–08 does not support the notion that the EU ETS would wreck the overall economy (as some claimed before the start of the scheme). On the contrary, it seems to indicate that an economy can grow moderately, even if key parts of its activities are faced with capped emissions.

Industrial competitiveness and carbon leakage: definitions
In the context of climate change policies, the term industrial competitive-ness is usually defined at the sector level. It refers to the ability of a sector in a certain country or region to maintain its profits and market shares vis-à-vis a similar sector in another country or region (Reinaud 2008). Within this context, the issue of industrial competitiveness refers usually to the problem that some countries accept and implement GHG abatement poli-cies, while others do not. Consequently, firms and sectors from nonabat-ing parties enjoy a comparative advantage, as they are not faced by costs or other constraints due to GHG mitigation.

The term carbon leakage refers to the increase in CO_2 (and other GHG) emissions in nonabating countries resulting from the mitigation actions in abating countries, thereby reducing the effectiveness of these actions. More precisely, given the implementation of climate policy in CO_2 abating country A and the resulting rise in CO_2 emissions in nonabating country NA, carbon leakage is usually defined as the ratio between the policy-induced increase of emissions in country NA and the reduction of emissions in country A. For instance if country A implements measures to reduce its emissions by 10 $mtCO_2$ while emissions in country NA increase by 2 $mtCO_2$ due to these measures, carbon leakage is equal to 2/10 * 100 percent = 20 percent.

Uneven abatement policies among countries may lead not only to carbon leakage but also to "competitiveness leakage," defined as a mitigation-induced shift in competitiveness—or comparative production and trade advantage—from abating to nonabating countries.

Identifying vulnerable industries
Several studies have tried to identify the sectors at risk of competitiveness/carbon leakage, based on an assessment of the major determinants of this risk.[18] In general, these determinants can be grouped into three sets: (1) factors affecting the exposure of industries to asymmetric increases in carbon costs, (2) factors affecting the ability to pass through asymmetric increases, and (3) other factors affecting industrial competitiveness and carbon leakage. These three sets of factors correspond to a three stage process used in a study by Climate Strategies (see Hourcade et al. 2007) to assess the potential impact of the EU ETS on industrial competitiveness and, more particularly, to identify U.K. industries at risk of carbon leakage. Both these factors and the three stages are discussed below.

Exposure of industries to asymmetric increases in carbon costs The extent to which industries are at risk of "competitiveness" carbon leakage depends first of all on the impact of asymmetric climate policies on their production costs. In turn, this impact depends mainly on the following.

- The energy or carbon intensity of the output produced. In addition to the direct cost increases, this factor refers also to the indirect cost impact of climate-policy-induced increases in electricity prices, notably for power intensive sectors such as the aluminum or copper industries.
- The type of climate policy (energy or carbon tax, emissions trading, energy efficiency regulation), as well as the specifics of this policy, including exemptions, free allocation conditions, compensatory measures, and so on.
- The stringency of the climate policy, which in the case of CO_2 emissions trading, for example, is a major determinant of the cost per ton of carbon.

In order to assess the cost exposure of industries to climate policy, different indicators or measures can be used: for instance, by expressing energy or carbon costs of specific industries as a percentage of their total sales revenues, production costs, net earnings, or profits (Carbon Trust 2004; Stern 2006; Houser et al. 2008; de Bruyn et al. 2008).

An alternative indicator is the so-called "value-at-stake" measure used by Hourcade et al. (2007) in their Climate Strategies study on the competitiveness impact of the EU ETS. This measure is defined as

value-at-stake = increase in total costs after allowance allocation divided by gross value added (GVA)

GVA = value of goods and services produced minus costs of raw materials and other inputs.

In order to capture both the direct and indirect cost aspects of the EU ETS, the value-at-stake concept is distinguished into

net value-at-stake (NVAS) = indirect cost impact due to EU-ETS-induced increases in electricity prices relative to GVA

maximum value-at-stake (MVAS) = direct and indirect cost impact of EU ETS relative to GVA, based on full purchasing of E.U. allowances by firms.

To estimate the value-at-stake impacts, the Climate Strategies study assumes a carbon price of 20 euros per tCO_2 and an induced electricity price increase of 10 euros per MWh. In the first stage of determining which sectors are at risk of carbon leakage, the study uses a threshold of 2 percent for NVAS and 4 percent for MVAS. (In other words, industries for which the NVAS is greater than 2 percent or the MVAS is greater than 4 percent are considered to be at risk of carbon leakage.) Using 2004 U.K. data for 159 manufacturing industries, only a few sectors exceeded the NVAS threshold of 2 percent—notably aluminum, fertilizers, nitrogen, and other inorganic basic chemicals—while 20 sectors recorded an MVAS greater than 4 percent, in particular industries producing cement, basic iron and steel, refined petroleum, or pulp and paper. Altogether, 23 sectors exceeded either the 2 or 4 percent threshold level. Direct emissions from these 23 sectors collectively contributed 11 percent of total U.K. GHG emissions, whereas their indirect emissions from electricity use contributed 3 percent. Their shares of U.K. GDP and employment were 1.1 and 0.5 percent, respectively (Hourcade et al. 2007).

Some qualifications, however, can be added to the above-mentioned results. First, the Climate Strategies study does not explain the choice of the threshold levels. One could argue that they are relatively low, but no clear objective way of identifying these levels is apparent.

Second, although the assumed carbon price of 20 euros per tCO_2 corresponds to the average E.U. allowance price in 2005–06, future carbon prices may actually be substantially higher. This implies that both the direct and indirect cost impacts of the EU ETS may become more significant and more relevant for a larger number of sectors exceeding critical threshold levels.

Third, the cost impact results depend on the level of (dis)aggregating industrial sectors. If the level of aggregation is relatively high, the (average) outcome for a rather heterogeneous sector may hide relevant differences in cost exposure to climate policy at a more disaggregated level. Moreover, even at a rather homogeneous or disaggregated sector level, certain intermediary products or parts of the production value chain may be traded or relocated individually. For instance, semifinished steel, clinker (input for cement), lime, basic glass and perhaps chemicals from steam crackers (ethylene, propylene, butane, and aromatics), ammonia, and pulp have the characteristics of high carbon intensity and relatively low value added, and they tend to be rather homogeneous products that either are internationally traded already or can be (Neuhoff and Dröge 2007). Hence, identifying industries at risk of carbon leakage has to be conducted at an appropriate disaggregated level.

Fourth, the cost impact results are based on 2004 U.K. data. Although similar results are available for comparable industrialized countries such as Germany (Hourcade et al. 2007), the United States (Houser et al. 2008), and the Netherlands (de Bruyn et al. 2008), these results may vary significantly over time—due to changes in market conditions and resulting output prices, affecting gross added values—as well as between countries depending on the structure and level of their industrial development.

Finally, although the share of the 23 sectors at risk of carbon leakage is relatively small in terms of national GDP or employment, they are generally far more important in terms of socioeconomic emanation or political sensitivity at the regional or local level.

Ability to pass through asymmetric increases in carbon costs   Another factor relevant to identifying industries at risk of carbon leakage is their ability to pass asymmetric, abatement-induced cost increases through to output prices. A simple but popular indicator for this pass-through ability is the international trade exposure or trade intensity of industries. This is based on the assumption that sectors with significant volumes of imports from or exports to countries outside the area with high carbon costs are likely not to pass all these costs through to output prices.

In the Climate Strategies study on the EU ETS, the ability of U.K. industries to pass through ETS-induced increases in (direct and indirect) carbon costs is identified by means of the so-called "non-E.U. trade intensity" measure (Hourcade et al. 2007). This measure is defined as follows.[19]

$$non\text{-}E.U.\ trade\ intensity = \frac{value\ of\ exports\ to\ non\text{-}E.U.\ +\ value\ of\ imports\ from\ non\text{-}E.U.}{annual\ turnover\ +\ value\ of\ imports\ from\ E.U.\ +\ value\ of\ imports\ from\ non\text{-}E.U.}$$

Based on 2004 data, the U.K. trade intensity outside the European Union varied from 0 percent for the power sector to 20–30 percent for refined petroleum and basic metals (including iron and steel) and even 40–50 percent for textiles and nonferrous metals (including aluminum and copper).

By combining the two metrics on cost and trade exposure, a first quantitative overview can be obtained of which sectors may be at risk, for instance by plotting cost exposure on the y-axis of a chart and trade exposure on the x-axis (see Hourcade et al. 2007 for the United Kingdom or Houser et al. 2008 for the United States). For the U.K. study on the EU ETS impact on industrial competitiveness and carbon leakage, this approach shows that, for some activities, the metrics on both "value-at-stake" and "trade intensity outside the European Union" are relatively high, in particular for basic metals, nonferrous metals, coke ovens, and refined petroleum. For other sectors, however, the cost exposure due to the EU ETS is relatively high whereas the trade exposure is relatively low (or vice versa), notably for the power sector (zero trade exposure) and the cement, lime, and plaster industry (about 5 percent trade intensity).

Besides the qualifications outlined above with regard to the cost impact metric, some further remarks can be added to the use of the trade exposure measure. First, trade intensity is an imperfect indicator for the ability of sectors to pass on carbon costs to output prices, because trade exposure is a dynamic parameter that may vary significantly between countries but can change substantially over time in response to price changes. For instance, while the level of steel traded outside the European Union is insignificant for Germany, it represents a large share of the U.K. market (Neuhoff and Dröge 2007). As noted, however, these trade exposure figures may change substantially within a decade.

In addition, the ability to pass through cost increases depends not only on exposure to international trade but also to the structure of the market. This refers particularly to (1) the number of firms active in a market (as an indicator for the level of market concentration or market competition) and

(2) the responsiveness of market demand to price changes of own products or substitutes (Sijm et al. 2008a; Sijm et al. 2008b; Sijm et al. 2009). Hence, firms in less competitive markets with low demand responsiveness may largely maintain sales volumes, market shares, and business profits, even if they are faced by asymmetric cost increases and exposure to outside trade. On the other hand, producers in unexposed or protected sectors may lose sales volumes and/or business profits due to high demand responsiveness to carbon cost-induced price increases, with demand—and related emissions—partially leaking to other sectors.

Finally, the impact of carbon abatement policies on industrial competitiveness and carbon leakage depends not only on simple, quantitative measures such as cost or trade exposure, but also on a variety of other, less quantifiable factors.

Other factors affecting industrial competitiveness and carbon leakage　As noted, in addition to the exposure of industries to asymmetric increases in carbon costs as well as their ability to pass these increases through to output prices, a variety of other factors affect industrial competitiveness and carbon leakage. In general, these factors refer to a variety of barriers to trade and (re)location of production, including the following in particular.[20]

Transport costs　Transport costs may act as a barrier to trade and hence to carbon leakage, depending on characteristics such as geographical location, mode of transport, bulkiness, or value added of the goods produced. For instance, cement is a relatively bulky, low value good. As a result, transporting cement by road is rather expensive, while it is much cheaper by international shipping. Therefore, whereas unilateral climate policies may have an adverse impact on the competitiveness of cement industries near to international shipping facilities, they may hardly affect the competitiveness of more inland cement industries, owing to the protection resulting from relatively high transport costs (Demailly and Quirion 2006).

Transport hazard　The production of chlorine is a relatively power-intensive activity and hence faces high indirect cost increases due to climate policy (just like aluminum or copper). Chlorine, however, is a very hazardous substance, which might restrict its scope for transport and, therefore, its risk of carbon leakage (Neuhoff and Dröge 2007).

Abatement and innovation potential　The vulnerability of firms and sectors to the cost of mitigation policies and hence to the risk of carbon leakage depends also on their abatement potential or, more generally, on their innovation potential to produce less carbon-intensive goods and services.

Trade restrictions   Import tariffs, export duties, technology standards, product labelling, and health or other quality controls may all act as barriers to trade and hence limit the risk of carbon leakage. On the other hand, in order to circumvent trade restrictions such as import tariffs or export duties, firms may decide to (re)locate production into domestic markets. Hence, the incidence of such trade restrictions enhances the risk of carbon leakage via the trade and relocation of production factors (the "investment channel"), while reducing the risk of carbon leakage through the trade of goods (the "output channel").

Product and service differentiation   A major strategy of companies within a certain sector, notably in more developed countries, is product differentiation by offering specialized, more sophisticated, or high quality commodities—including brand names—that meet the specific demand of certain industries or end-users. A related strategy is service differentiation, including certainty in product availability and time of delivery, price stability, quality control, information, support, maintenance, and so on. In general, such product or service differentiation reduces competition and enhances price margins, thereby lowering the risk of carbon leakage (Hourcade et al. 2007).

Complex, capital-intensive investments   The production of carbon-intensive goods such as steel, cement, chemicals, and refined oil products usually requires complex, high capital investments in facilities lasting for several decades. The costs of these investments are covered in years when scarce production capacity results in scarcity premiums (Neuhoff and Dröge 2007). Therefore, as producers in such industries are used to taking long-term perspectives on investment and operational decisions, this reduces the risk of carbon leakage in the short or medium term, while in the long run this risk may be reduced due to the opportunity of multilateral climate policies equalizing the global playing field.

Other trade and relocation barriers   In addition, a variety of other trade and relocation barriers limit the risk of carbon leakage. These include production or investment determinants such as proximity to markets, natural resource input availability, labor costs, quality of human resources, political risks, macroeconomic and social stability, adequate legal regimes (for example, intellectual property rights, contract law, investment law, an independent judiciary), infrastructure (communications, energy, transportation), and other considerations (Cosbey and Tarasofsky 2007).

Note that the incidence and significance of the trade and relocation barriers outlined above may vary between countries and industries. Therefore, even if industries in abating countries are faced by similar exposures to international trade intensities and mitigation-induced cost increases, the risk of carbon leakage may vary significantly between these countries and industries depending on the incidence and importance of these barriers.

The incidence of carbon leakage due to the EU ETS
Several (modeling) studies have tried to assess the impact of the EU ETS on industrial competitiveness and carbon leakage.[21] In general, however, the findings of these studies vary widely, depending on the sectors considered and the data, methodology, and assumptions used. For instance, at a carbon price of 20 euros per tCO_2 in the EU ETS, model estimates of carbon leakage range from 0.5 to 25 percent in the iron and steel sector and from 40 to 70 percent in the cement sector, depending on how allowances are allocated among other parameters (Demailly and Quirion 2006 and 2008; Ponssard and Walker 2008).

Empirically, however, there is no evidence of significant carbon leakage for the sectors concerned during the first phase of the EU ETS (2005–07). Apart from the overall favorable world economic conditions during the years 2005–07 and the, in general, generous allocations of free allowances to these sectors (including the related plant closure conditions to these allocations), this is probably also due to the relatively short period considered, which does not allow observation of the full potential long-term effects of the EU ETS on industrial competitiveness and carbon leakage (Reinaud 2008; Convery et al. 2008). Hence, any impact of the EU ETS on the performance of industrial sectors is likely to become more significant when markets are less favorable, carbon prices are higher, and/or allocations of allowances to industries are less generous.

CHANGES IN THE EU ETS BEYOND 2012

In January 2008 the European Commission proposed an energy and climate policy package for the period up to 2020 and beyond. The two key objectives of this package are (1) to reduce overall GHG emissions to 20 percent below 1990 levels by 2020 (possibly scaling up to 30 percent in the event of a satisfactory international agreement being reached) and (2) to increase the share of renewable energy sources to 20 percent by 2020. In December 2008 an amended version of this package was adopted by the

European Council of Ministers, representing the member states, and the European Parliament. A core element of the policy package is a major revision and strengthening of the EU ETS, starting from 2013 up to 2020 and beyond. The major changes for the EU ETS after 2012 include the following (European Commission 2009; World Bank 2009).

A more stringent, single E.U.-wide cap   Sectors and activities covered by the EU ETS have to reduce their emissions by 21 percent below 2005 levels. Starting from 2013, a single E.U.-wide cap will be set centrally by the European Commission. For sectors included in the ETS, the cap on emissions is expected to decrease at a rate of 1.74 percent per year with the 2010 allocation as a reference. Based on phase 2 coverage and allocation (2,080 million EUAs per year, on average), this would correspond to an E.U.-wide allocation of 1,974 million EUAs by 2013, decreasing to 1,720 million EUAs by 2020.

Harmonized allocation rules   Besides a single E.U.-wide cap, other elements of harmonized allocation include a sole E.U.-wide new entrants reserve (5 percent of the entire amount of allowances) and centralized rules for auctioning and free allocations to installations.

Auctioning   Starting from 2013, about half of all allowances will be auctioned, increasing with time until 70–80 percent of the allowances are auctioned by 2020. Allowances are to be auctioned by member states, with national shares largely reflecting phase 1 emissions.

Auctioning for electricity producers   Full auctioning will start in 2013 for power producers, with concessions made to some member states, taking into account the status of their electricity sector and GDP per capita. For existing installations, these member states will have the option to start auctioning at least at 30 percent by 2013 reaching 100 percent by 2020.

Free allocation and phased auctioning for industry and other sectors   E.U.-wide rules for free allocation were scheduled for adoption by 31 December 2010 with the intent of harmonizing these rules across member states. For industries not exposed to global competition, auctioning will be phased in gradually, starting with a modest 20 percent in 2013 and increasing to 70 percent by 2020 (with a view to finally reaching full auctioning by 2027). For those sectors exposed to global competition, the aggregate number of free allowances for this group will be set in proportion to their historical share of emissions during phase 1 and will decline annually in proportion to the overall phase 3 cap. Free allocation to individual installations in both industry categories will, "to the extent possible," be based on bench-marking to the best available technology. The intent is that free allocation rewards efficient installations more than less efficient installations in any sector.

The sectors and subsectors exposed to global competition (and those that are not), were to be determined, based on an assessment of projected increases in production costs as a result of carbon regulation and degree of openness. The exposure of installations to international competition was to be assessed in depth, and additional measures to protect these industries could be proposed as needed (World Bank 2009).

Coverage   Aviation will already be included in the EU ETS starting from 2012. The next year, the scope of the scheme will be further extended by covering CO_2 emissions from petrochemicals, ammonia, and aluminum, N_2O emissions from nitric, adipic, and glyocalic acid production, and per-fluorocarbons from the aluminum sector.

Trading period   The third trading period will last eight years, from 2013 up to 2020.

SUMMARY OF EU ETS EXPERIENCES, ACHIEVEMENTS, AND LESSONS LEARNED

This section provides a bullet-point summary of the major experiences, achievements and lessons learned by the EU ETS during the first three to five years of its existence under five theme headings.[22]

Allocation system

- Reliable emissions data and good projections at the installation level are essential for (avoiding over-) allocation of free allowances.
- The system of decentralized and free allocations of emission allowances during the first and second phase of the EU ETS facilitated acceptance and introduction of the scheme.
- On the other hand, this system of decentralized and free allocation has caused a variety of inefficiencies, competitive distortions, and equity problems, which threaten the long-term cost-effectiveness, integrity, and acceptability of the scheme.
- Phase 1 allocation was a useful experience: some findings and lessons learned during this trial period were used to improve the performance

of the allocation system, albeit modestly for the second phase, but more fundamentally for the third phase and beyond through a decisive revision of the EU ETS directive for the period after 2012.

Carbon abatement

- A few studies indicate that some abatement of CO_2 emissions occurred during the first phase of the EU ETS, despite an overall modest cap and widespread over-allocations of allowances at the installation, sector, and country levels.
- Besides some small energy efficiency improvements across the scheme (induced by relatively high EUA prices in 2005–06), carbon abatement during the first phase of the scheme was most likely restricted to short-term fuel switching in the power sector of some countries: notably Germany and the United Kingdom.
- Emission reductions appear occasionally where they are not expected—for example, through switching from lignite to coal or from coal to gas or biomass as in Germany, or through improvement of coal generation by increasing biomass use or enhancing energy efficiency as in the United Kingdom (Convery et al. 2008).

EUA market development

- Since 2005, the E.U. allowance market has developed strongly, resulting in some major achievements, in particular (1) the EUA market infrastructure (that is, trading platforms, registries, and so on) is in place; (2) the EUA market has expanded rapidly in terms of both volume and value of transactions; (3) the EUA market has strongly encouraged the development and growth of the JI and CDM market (through the linkage or use of JI and CDM credits to meet EU ETS compliance); (4) the EUA market has established one single E.U.-wide, transparent carbon price; and, perhaps most importantly, (5) the EUA market made stakeholders aware that carbon emissions have a price to be internalized in their decisionmaking.
- Timely and reliable information on the verified emissions and (long-term) EU ETS cap, as well as opportunities for banking (and borrowing) of allowances within and between (adequately long) trading periods are essential for market efficiency, EUA price stability, and proper investment decisions affecting carbon emissions.

Impact on growth, competitiveness, and carbon leakage

- For the years 2005–08, there are no significant empirical signs that the EU ETS has exerted an adverse effect on economic growth or industrial competitiveness of participating sectors, or that it resulted in carbon leakage to other countries.
- The impact of the EU ETS on economic growth, industrial competitiveness and carbon leakage, however, may be larger in the long run when EUA prices are higher (while outside competitors do not face similar costs), when world markets are less buoyant (compared with 2005–07), or when free allocation of allowances (including closure conditions) are less favorable to incumbents and new entrants.

EU ETS overall performance

- The introduction of a short pilot phase was useful: lessons learned were, to some extent, already applied to phase 2 and, more importantly, largely incorporated in the drastic revisions of the EU ETS directive for phase 3 and beyond.
- Despite several problems, shortcomings, and poor conditions during the trial phase, the system performed surprisingly well during the first years of its operation, in particular in terms of EUA market development, raising carbon cost awareness and internalizing EUA pricing in stakeholders' operational decisions (Ellerman and Joskow 2008).
- Up to now, the performance of the EU ETS has been low (or largely unknown) in terms of encouraging R&D and implementation of new, carbon-saving technologies.
- Over the five years 2005–10, the EU ETS became both the cornerstone and the flagship of E.U. climate policy.
- A major lesson of the first years of the EU ETS is that not everything has to be perfect to get started: although imperfections were reflected in EUA price volatility—or other market inefficiencies—they did not really hinder the development of the EUA market, whereas the existence of this market was actually the best stimulus to address these imperfections.
- The EU ETS experience with the so-called "dynamics of harmonization versus differentiation" (that is, centralized versus decentralized policy control and implementation) offers useful lessons for other regional and global ET systems or mitigation policies (Kruger et al. 2007; Ellerman 2008).

- Perhaps one of the main "achievements" of the EU ETS is that it reflects and establishes a "cultural change." While during the Kyoto protocol negotiations the European Union was still opposed to emissions trading, within a decade both the idea and practice of emissions trading—and market-based environmental policy instruments in general—were widely and increasingly accepted within the European Union.
- Overall, despite a variety of shortcomings of the EU ETS up to now, the European Commission—in particular the officials of the Directorate-General for the Environment designing, implementing, and revising the EU ETS directive—did an amazingly good job.

ACKNOWLEDGMENTS

A first version of this study was presented at the Conference on Climate Change and Green Growth: Korea's National Growth Strategy (Honolulu, Hawaii, 23–24 July 2009), organized by the East-West Center (EWC) in Hawaii and the Korea Development Institute (KDI). The author would like to thank the organizers of this conference for the invitation and support to write and present this paper. In addition, he would like to thank Dr. Taehoon Yoon (KDI) and other participants at the conference for comments on the first version of this paper. Finally, the author would like to express his gratitude for additional support offered by his employer, the Energy Research Centre of the Netherlands, where the work and support for this paper are registered under project number 50.250.01.03.

NOTES

1. Some interesting studies and papers have evaluated the performance of the EU ETS during its pilot phase (2005–07). See in particular Convery et al. (2008), Ellerman and Joskow (2008), and Ellerman et al. (2007). The present chapter builds on these publications as well as on the author's own publications and other references mentioned in the chapter.
2. When the Kyoto protocol was agreed (1997), the European Union consisted of fifteen member states (the EU-15). Since then, the European Union has expanded to its present number of twenty-seven member states (the EU-27), including several countries in Central and Eastern Europe.
3. In the EU ETS, a distinction is made between an *allowance* and a *permit*. An EU ETS allowance gives the right to emit one tonne of CO_2. It is the system's tradable unit, called the "E.U. allowance" (EUA). On the other hand, each installation in the EU ETS must have a permit from its competent authority. It is a kind of (nontradable) license, which sets certain conditions on an installation's operation, in particular that the operator is capable of monitoring and reporting the plant's emissions.

4. In addition, during the first phase, member states were allowed to opt-in installations below capacity limits in EU ETS sectors, while from 2008 on they could also opt-in certain other activities, installations, and GHGs (notably facilities in the chemical industry emitting N_2O). In general, however, the size of these options was relatively small in terms of total GHG emissions.
5. For more details on the EU ETS allocation system up to 2012 in general and the NAPs during the first and second trading periods in particular, see Gilbert et al. (2006), Neuhoff, Åhman, et al. (2006), Ellerman et al. (2007), and the special website of the European Commission at ec.europa.eu/environment/climat/campaign/index_en.htm.
6. The Commission made some major adjustments to the draft NAPs proposed by the member states. Overall, the Commission reduced the amount of CO_2 allowances in fifteen NAPs for the first allocation period by, in total, 290 mt annually and in twenty-three NAPs for the second phase by 242 mt annually. In addition, the Commission limited the number of JI and CDM credits that participants could import into the ETS during phase 2, while it rejected any ex-post allocation adjustments—and some other specific allocation provisions—in NAPs I and II (Convery et al. 2008).
7. During the first phase of the EU ETS, only four small member states auctioned small percentages of their allowances: Denmark (5 percent), Hungary (2.5 percent), Lithuania (1.5 percent), and Ireland (0.75 percent). During the second phase, auctioning was used by Germany (8.8 percent), the United Kingdom (7.0 percent), the Netherlands (4 percent), Lithuania (2.9 percent), Hungary (2.3 percent), Austria (1.2 percent), Ireland (0.5 percent), and Belgium (0.3 percent). Overall, the average of total E.U. allowances auctioned amounted to only 0.13 percent in phase 1 and 3.0 percent in phase 2 (Ellerman and Joskow 2008).
8. See Clo (2009). His findings, however, are not based on (the equalization of) marginal abatement cost between ETS and non-ETS sectors, but rather on two benchmarks or "proportional Kyoto targets," determined by multiplying, for any member state, its Kyoto target by the pre-2005 and 2005 ETS share in its total emissions, respectively.
9. An alternative and simpler definition of over-allocation could use a standard in which the reference emissions are determined at a certain fixed amount, say minus 10 percent below a historic baseline level of emissions. Such a definition would avoid the construction of a counterfactual estimate of EU ETS emissions (although such a counterfactual would still be necessary to estimate the level of abatement due to the ETS).
10. For similar analyses and findings, see Ellerman and Buchner (2008).
11. For the first phase as a whole (2005–07), the net long positions amounted to some 50 million tons of CO_2 per year on average, or about 2.3 percent of the total allocated allowances over this period (Anderson and di Maria 2009). As this surplus of allowances during the first phase could not be banked for subsequent trading periods, it implied that these allowances became actually worthless and hence ready to be destroyed.
12. As a percentage of their allocated allowances, these amounts correspond to 13.3 percent (Poland), 5.0 percent (Germany), 13.1 percent (France), and 14.4 percent (the Czech Republic), respectively (Kettner et al. 2008).
13. Convery et al. (2008) report preliminary results from more focused research on the German power sector, which support this finding of moderate abatement. More specifically, a shift in generation from higher emitting lignite (brown coal) to lower emitting hard coal can be observed, as well as an increase in the use of biomass. Also, in the United Kingdom, more focused research indicates a noticeable improvement in the CO_2 efficiency of coal-fired generating plants. This could have been due to increased use of biomass or improved energy efficiency in response to the sharp increase in the cost of using coal to generate electricity (Convery et al. 2008).
14. Anderson and di Maria (2009) define emissions inflation as

> behaviour that leads to higher emissions levels than what would have occurred in the absence of the trading scheme, that is emissions greater than the business as usual levels. This is possible and likely in the context of the EU ETS due to the

methodology used for pilot phase allocation and uncertainty about future allocation methodologies. In the pilot phase, most governments allocated total emissions
relative to 'business-as-usual' projections, and the more detailed distribution [of
allowances] has typically occurred in relation to past emissions (Grubb et al. 2005).
EU ETS participants may have learned that inflating (historical) emissions leads
to more generous future allocations. Grubb et al. (2005) point out that emissions
inflation due to the prospect of future allowance distribution being contingent upon
recent emissions ('updating') is likely, and gives a direct incentive to industries to
inflate actual emissions.

15. Due to the severe economic crisis during the first years of the second trading period,
    however, a study by Sandbag (2009) estimates that overall a total surplus of 700 mtCO$_2$
    emission allowances could be available in phase 2 of the scheme, which are then bank-
    able for use up to 2020. Including JI and CDM credits, there could even be a surplus
    available of 1.6 billion mtCO$_2$ emission allowances and credits, all bankable for use in
    the future.
16. As mentioned in the second section above, these registries, in which ETS installations
    must open accounts, are organized by member states in order to register the allowance
    allocations to these installations and track all movements of allowances resulting from
    market or compliance transactions.
17. Several studies have analyzed the determinants of the EUA price (or the stochastic
    behavior of this price). See among others Alberola et al. (2007 and 2008) and Chevalier
    (2009). For recent overviews of these studies, see Bonacina and Cozialpi (2009) and
    Bonacina et al. (2009).
18. See, in particular, Carbon Trust (2004 and 2006), Stern (2006), McKinsey and Ecofys
    (2006), Hourcade et al. (2007), Houser et al. (2008), and de Bruyn et al. (2008).
19. In addition, the CS study applies the "E.U. trade intensity" measure in order to
    account for the trade exposure to other E.U. countries. Whereas the "non-E.U. trade
    intensity" measure can be considered primarily as an indicator for the ability to pass
    through ETS-induced cost increases and, hence, for the risk of competitiveness/carbon
    leakage due to the nonabating party problem, the "E.U. trade intensity" measure
    could be regarded as an indicator for the impact on industrial competitiveness within
    the European Union due to the differentiated implementation problem, notably the
    problem of differential allocation methods and volumes between member states (as
    discussed in the third section of this chapter).
20. See Hourcade et al. (2007) who, in the third stage of their process to assess the impact of
    the EU ETS on industrial competitiveness, conduct a deep-dive study in the cement and
    steel sectors in order to explore these other factors. See also Neuhoff and Dröge (2007),
    Cosbey and Tarasofsky (2007), and Reinaud (2008).
21. See Reinaud (2008) for references and a review of these studies.
22. Similar and other experiences, achievements, and lessons learned by the EU ETS have
    been discussed and reviewed by, in particular, Convery et al. (2008), Ellerman and
    Joskow (2008), Convery (2008), Fujiwara and Egenhofer (2008), EAC (2007), Ellerman
    et al. (2007), Kruger et al. (2007), Buchner et al. (2006), and Betz and Sato (2006).

REFERENCES

Åhman, M., and K. Holmgren. 2006. New Entrant Allocation in the Nordic
    Energy Sectors: Incentives and Options in the EU ETS. *Climate Policy* 6 (4):
    423–40.
Alberola, E., J. Chevalier, and B. Chèze. 2007. *European Carbon Prices
    Fundamentals in 2005–2007: The Effects of Energy Markets, Temperatures and*

Sectorial Production. Economics Working Paper 2007-33. Paris: Université Paris X Nanterre.

Alberola, E., J. Chevalier, and B. Chèze. 2008. Price Drivers and Structural Breaks in European Carbon Prices 2005–2007. *Energy Policy* 36:787–97.

Anderson, B., and C. di Maria. 2009. Abatement and Allocation in the Pilot Phase of the EU ETS. Paper presented at the Seventeenth Annual Conference of the European Association of Environmental Resource Economics (EAERE), Amsterdam, 24–27 June 2009.

Betz, R., and M. Sato. 2006. Emissions Trading: Lessons Learnt from the 1st Phase of the EU ETS and Prospects for the 2nd Phase. *Climate Policy* 6:351–59.

Bonacina, M., and S. Cozialpi. 2009. *Carbon Allowances as Inputs or Financial Assets: Lessons Learned from the Pilot Phase of the EU ETS.* Working Paper 19. Milan: IEFE (Centre for Research on Energy and Environmental Economics and Policy), Universitá Commerciale Luigi Bocconi.

Bonacina, M., A. Creti, and S. Cozialpi. 2009. *The European Carbon Market in the Financial Turmoil: Some Empirics in Early Phase II.* Working Paper 20. Milan: IEFE (Centre for Research on Energy and Environmental Economics and Policy), Universitá Commerciale Luigi Bocconi.

Bruyn, S. de, D. Nelissen, M. Korteland, M. Davidson, J. Faber, and G. van de Vreede. 2008. *Impacts on Competitiveness from EU ETS: An Analysis of the Dutch Industry.* Delft: CE Delft.

Buchner, B., C. Carraro, and D. Ellerman. 2006. *The Allocation of European Union Allowances: Lessons, Unifying Themes and General Principles.* FEEM Working Paper 116.06. Venice: University of Venice.

Carbon Trust. 2004. *The European Emissions Trading Scheme: Implications for Industrial Competitiveness.* London: Carbon Trust.

Carbon Trust. 2006. *Allocation and Competitiveness in the EU Emissions Trading Scheme.* London: Carbon Trust.

Chevalier, J. 2009. Carbon Futures and Macroeconomic Risk Factors: A View from the EU ETS. *Energy Economics* 31:614–25.

Clo, F. 2009. The Effectiveness of the EU Emissions Trading Scheme. *Climate Policy* 9:227–41.

Convery, F. 2008. Reflections on the European Union Emissions Trading Scheme (EU ETS): What Can We Learn? Paper presented at the Informational Board Workshop on Policy Tools for the AB 32 Scoping Plan, Air Resources Board, California Environmental Protection Agency, Sacramento, California, 28 May 2008.

Convery, F., D. Ellerman, and C. de Perthuis. 2008. *The European Carbon Market in Action: Lessons from the First Trading Period.* Interim Report, MIT Joint Program on the Science and Policy of Global Change, Report 162. Boston, Massachusetts: Massachusetts Institute of Technology.

Cosbey, A., and R. Tarasofsky. 2007. *Climate Change, Competitiveness and Trade.* London: Chatham House.

Delarue, E., D. Ellerman, and W. D'haeseleer. 2008. *Short-term CO_2 Abatement in the European Power Sector.* MIT Center for Energy and Environmental Policy Research, Working Paper 0808. Leuven: University of Leuven; Boston, Massachusetts: Massachusetts Institute of Technology.

Delarue, E., K. Voorspoels, and W. D'haeseleer. 2008. Fuel Switching in the Electricity Sector in the EU ETS: Review and Prospective. *Journal of Energy Engineering* 134 (2): 40–46.

Delbeke, J., G. Klaassen, T. van Ierland, and P. Zapfel. 2009. The Economics of the New EU Climate and Energy Package. Paper presented at the Seventeenth Annual Conference of the European Association of Environmental Resource Economics (EAERE), Amsterdam, 24–27 June 2009.

Demailly, D., and P. Quirion. 2006. CO_2 Abatement, Competitiveness and Leakage in the European Cement Industry under the EU ETS: Grandfathering versus Output-based Allocation. *Climate Policy* 6:93–113.

Demailly, D., and P. Quirion. 2008. European Emission Trading Scheme and Competitiveness: A Case Study on the Iron and Steel Industry. *Energy Economics* 30:2009–27.

Ellerman, D. 2008. *The EU's Emissions Trading Scheme: A Proto-type Global System?* CEEPR Working Paper 08-013. Cambridge, Massachusetts: Center for Energy and Environmental Policy Research, Massachusetts Institute of Technology.

Ellerman, D., and B. Buchner. 2007. The European Union Emissions Trading Scheme: Origins, Allocation, and Early Results. *Review of Environmental Economics and Policy* 1 (1): 66–87.

Ellerman, D., and B. Buchner. 2008. Over-Allocation or Abatement? A Preliminary Analysis of the EU ETS Based on the 2005–06 Emissions Data. *Environmental Resource Economics* 41:267–87.

Ellerman, D., and S. Feilhauer. 2008. *A Top-down and Bottom-up Look at Emissions Abatement in Germany in Response to the EU ETS.* CEEPR Working Paper 08-017. Cambridge, Massachusetts: Center for Energy and Environmental Policy Research, Massachusetts Institute of Technology.

Ellerman, D., and P. Joskow. 2008. *The European Union's Emissions Trading System in Perspective.* Arlington, Virginia: Pew Center on Global Climate Change.

Ellerman, D., B. Buchner, and C. Carraro. 2007. *Allocations in the European Emissions Trading Scheme: Rights, Rents and Fairness.* Cambridge: Cambridge University Press.

Environmental Audit Committee (EAC). 2007. *The EU Emissions Trading Scheme: Lessons for the Future.* Report by the Environmental Audit Committee of the House of Commons. London: Environmental Audit Committee.

European Commission (EC). 2003. Directive 2003/87/EC of the European Parliament and of the Council of 13 October 2003 Establishing a Scheme for Greenhouse Gas Emission Allowance Trading within the Community and Amending Council Directive 96/61/EC. *Official Journal of the European Union* 25.10.2003: L275/32-46.

European Commission. 2007. *EU Action against Climate Change—EU Emissions Trading: An Open System Promoting Global Innovation.* Brussels: European Commission.

European Commission. 2009. Directive 2009/29/EC of the European Parliament and of the Council of 23 April 2009 Amending Directive 3003/87/EC so as to Improve and Extend the Greenhouse Gas Emission Allowance Trading Scheme of the Community. *Official Journal of the European Union* 5.6.2009: L140/63-87.

Eurostat. 2009. *Statistics Database.* Available at http://epp.eurostat.ec.europa.eu.

Fujiwara, N., and C. Egenhofer. 2008. *What Lessons Can Be Learned from the EU Emissions Trading Scheme?* CEPS Policy Brief 153. Brussels: Centre for European Policy Studies.

Gilbert, A., G. Reece, D. Phylipsen, K. Mirowska, M. Horstink, and J. Stroet.

2006. *Comparative Analysis of National Allocation Plans for Phase I of the EU ETS*. Final report to the Government of Ireland, YUES5023. London: Ecofys UK.

Grubb, M., C. Azar, and M. Persson. 2005. Allowance Allocation in the European Emissions Trading System: A Commentary. *Climate Policy* 5:127–36.

Harrison, D., P. Klevnas, D. Radov, and A. Foss. 2007. *Complexities of Allocation Choices in a Greenhouse Gas Emissions Trading Program*. Report to the International Emissions Trading Association (IETA). Boston, Massachusetts: NERA Economic Consulting.

Hourcade, J., D. Demailly, K. Neuhoff, and M. Sato. 2007. *Differentiation and Dynamics of EU ETS Industrial Competitiveness Impacts*. London: Climate Strategies.

Houser, T., R. Bradley, B. Childs, J. Werksman, and R. Heilmayr. 2008. *Leveling the Carbon Playing Field: International Competition and US Climate Policy Design*. Washington: Peterson Institute for International Economics, World Resources Institute.

Kettner, C., A. Köppl, S. Schleicher, and G. Thenius. 2008. Stringency and Distribution in the EU Emissions Trading Scheme: First Evidence. *Climate Policy* 8 (1): 41–61.

Kruger, J., W. Oates, and W. Pizer. 2007. Decentralization in the EU Emissions Trading Scheme and Lessons for Global Policy. *Review of Environmental Economics and Policy* 1 (1): 112–33.

McKinsey and Ecofys. 2006. *EU ETS Review: Report on International Competitiveness*. Brussels: European Commission, Directorate General for the Environment.

Neuhoff, K., and S. Dröge. 2007. *International Strategies to Address Competitiveness Concerns*. Berlin: German Institute for International and Security Affairs (Stiftung Wissenschaft und Politik, SWP); Cambridge: Electricity Policy Research Group, University of Cambridge.

Neuhoff, K., M. Åhman, R. Betz, J. Cludius, F. Ferrario, K. Holmgren, G. Pal, M. Grubb, F. Matthes, K. Rogge, M. Sato, J. Schleich, J. Sijm, A. Türk, C. Kettner, and N. Walker. 2006. Implications of Announced Phase II National Allocation Plans for the EU ETS. *Climate Policy* 6 (4): 411–22.

Neuhoff, K., K. Keats, J. Sijm, F. Matthes, and A. Johnston. 2006. *Allocation Matters: So What Can We Do About It? Strategies for the Electricity Sector 2008–2012*. Study commissioned by Climate Strategies and The Carbon Trust, London.

Organisation for Economic Co-operation and Development (OECD). 2009. *Economic Outlook: Annual Projections for OECD Countries*. Paris: Organisation for Economic Co-operation and Development.

Ponssard, J., and N. Walker. 2008. EU Emissions Trading and the Cement Sector: A Spatial Competition Analysis. *Climate Policy* 8:467–93.

Reinaud, J. 2008. *Issues behind Competitiveness and Carbon Leakage: Focus on Heavy Industry*. Paris: International Energy Agency.

Sandbag. 2009. *ETS S.O.S.: Why the Flagship "EU Emissions Trading Policy" Needs Rescuing*. London: Sandbag Climate Campaign.

Sijm, J., K. Neuhoff, and Y. Chen. 2006. CO_2 Cost Pass-through and Windfall Profits in the Power Sector. *Climate Policy* 6 (1): 49–72.

Sijm, J., J. Hers, W. Lise, and B. Wetzelaer. 2008a. *The Impact of the EU ETS on Electricity Prices: Final report to the Directorate-General for the Environment*

(DG Environment) of the European Commission. ECN report ECN-E–08-007. Petten, Amsterdam: Energy Research Centre of the Netherlands (ECN).

Sijm, J., S. Hers, and B. Wetzelaer. 2008b. "Options to Address Concerns Regarding EU ETS-induced Increases in Power Prices and Generators' Profits: The Case of Carbon Cost Pass-through in Germany and the Netherlands." In *Markets for Carbon and Power Pricing in Europe: Theoretical Issues and Empirical Analyses*, edited by F. Gulli, pp. 101–44. Cheltenham, UK and Northampton, MA, USA: Edward Elgar.

Sijm, J., Y. Chen, and B. Hobbs. 2009. The Impact of Power Market Structure on the Pass-through of CO_2 Emissions Trading Costs to Electricity Prices: A Theoretical Approach. Paper presented at the Seventeenth Annual Conference of the European Association of Environmental Resource Economics (EAERE), Amsterdam, 24–27 June 2009.

Stern, N. 2006. *The Economics of Climate Change: Stern Review on the Economics of Climate Change.* London: Her Majesty's Treasury. Republished as *The Economics of Climate Change: The Stern Review* (Cambridge and New York: Cambridge University Press, 2007.)

Widerberg, A., and M. Wrake. 2009. *The Impact of the EU Emissions Trading Scheme on CO_2 Intensity in Electricity Generation.* Working Paper in Economics 361. Gothenburg: School of Business, Economics, and Law, University of Gothenburg.

World Bank. 2009. *State and Trends of the Carbon Market 2009.* Washington: World Bank.

4. Energy and climate change policy: perspectives from the International Energy Agency

Richard A. Bradley

INTRODUCTION

Balancing energy security with economic and climate change goals is one of the most difficult policy challenges facing member governments of the International Energy Agency (IEA). It is a challenge under normal circumstances; it seems nearly unachievable in the face of climate change. Emissions and concentrations are rising, impacts are already observable and ultimately may be quite significant, political consensus is elusive, and yet the policy action commensurate with the threat is urgent.

The key features of the policy challenge are the slow rate of capital stock turnover in the energy consuming and supply sector and the unprecedented geographic and economic sources of the emissions, which require both international and domestic policy formulation. Accordingly, climate change mitigation policy will be unique in its complexity, requiring a package of different measures—just as there is no technological silver bullet, there is no policy silver bullet—and the creation of new institutional forms.

No historic model exists to guide such complex policy formulation. This unprecedented global commons problem inspires no shortage of proposals for grand solutions, but the urgency for considerable action, along with the limits of governments to find the ideal within the complexity, argues for practical next steps. This chapter describes the key features of the mitigation challenge, and suggests some priorities for near-term response.

THE NATURE OF THE CHALLENGE

Climate change is a particularly difficult public-policy challenge because of inertia, both in the climate system, where greenhouse gases have such long atmospheric residence time, and in the existing capital stock. The long-lived nature of the gases makes climate change a stock externality and the act of emitting nearly irreversible. The Intergovernmental Panel on Climate Control (IPCC) has determined (Barker et al. 2007) that if concentrations are to be stabilized at levels below 600 parts per million (ppm), then it will be necessary to reduce global greenhouse gas (GHG) emissions to nearly net zero after accounting for GHG sinks. More challenging is that to reach those levels, we must reach the peak of global GHG emissions by 2030 at the latest. Most IEA countries have concluded that doubling preindustrial concentrations presents an unacceptable risk and have set concentration goals closer to 450–500 ppm, which would require reaching the apex of global emissions within the next decade (Barker et al. 2007; IEA 2008c).

If 450ppm is to remain an option, emissions must reach nearly net zero by the end of the century and must also peak by 2020. The investment challenge is that our current energy infrastructure must change rapidly, including infrastructure responsible for energy consumption. Such a change is in fact revolutionary, at least with respect to the GHG emissions profile of our capital structure.

It is not the creation of new capital *per se* that is revolutionary. After all, the energy infrastructure of 1950 was fundamentally different from the one that existed at the advent of the industrial revolution in 1850. Instead, much of the existing capital will, in contrast to that which existed in 1950, still exist for much of the next hundred years. In 1850, with the exception of buildings, little long-lived capital existed to contribute to the GDP of a largely agrarian economy. Now much of the energy consuming and supply capital structure plays a much larger role in the economy. Figure 4.1 shows the range of life spans for different types of capital. Buildings, transportation infrastructure, industrial facilities, and electric power generating plants all last for many decades and as long as a century. The challenge is that the change now requires replacement and modernization of capital that is still economically productive in almost all sectors of the economy over the same century timeframe.

Generally the capital in place has been expensed on the accounting books of firms. After being fully depreciated, the engineering performance of the capital may continue for decades, requiring only modest maintenance outlay while generating the same revenue stream. This is a period of considerable profitability for a technology, and foregoing that significant profit is what makes the opportunity cost of premature retirement so significant. As Blyth (2010) concludes, the "favourable economics of existing coal plant relies largely on the fact that the large capital costs have been sunk, and operating costs are typically relatively low, such that

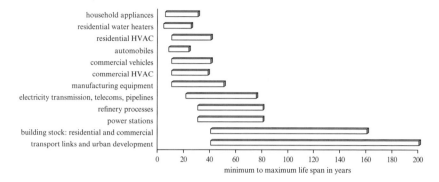

Source: Adapted from PNL/University of Maryland (2003).

Figure 4.1 Capital stock turnover rates: life spans by major type

they can make a return with relatively small profit margins." In reality, lifetimes of capital can be significantly longer than has been generally thought. For example, electricity generating plants may last up to 80 years, as illustrated by the turnover rates shown in Blythe's (2010) vintage profile of currently installed coal-fired plants in various countries.[1] This is consistent with the findings of a Pew Center study (Lampert et al. 2002) on U.S. power plants.

The implications of these long lifetimes for emissions can be considerable. These turnover rates mean that virtually all of the generating capacity constructed in 2010, for example, will still exist in 2030 (when global emissions must peak if ultimate concentrations below 600 ppm are to remain an option), and most of it will still have a useful, profitable life in 2050 in the absence of policies. In fact, most of the generating capacity constructed since 1975 will still be operating in 2030.

Blythe's figure shows for selected countries the age profile of their coal-fired generating capacity. Most new builds are in China, Australia, Canada, and Japan and in the "other" category of countries, while prior to 1983 most capacity additions were in the United States and Europe. The coal-fired capacity is older in the United States and Europe than in other large countries.

This has significance for the relative policy focus of countries. Countries with an older capital structure will need policies that influence both the extent and timing of refurbishment in addition to incentives to "green" new builds. Countries undergoing rapid industrialization will have a relative policy focus on incentives addressing new builds.

It is possible to modify the historic rates of capital turnover with policy, but at a cost, namely the loss of economically productive capital. The IPCC has shown (Barker et al. 2007) that the more stringent the target in terms of level or proximity of the target, the more expensive the transition, in part due to the premature retirement of capital. To the extent that we can make substantial changes in our energy supply and consumption infrastructure during capital's normal retirement or refurbishment cycles, the cost of meeting our climate goals can be lowered.

Geophysical and capital stock inertia is at the heart of the IEA's climate change policy recommendations. Overcoming the resistance to rapid change is a significant barrier to achieving the declared climate goals of IEA governments. Much of the resistance can be overcome by expending resources, but the nature and extent of the factors that inhibit rapid capital stock refurbishment and replacement are not clear. Much of the IEA's climate and energy security analysis is directed at identifying such factors and developing policy options.

A POSSIBLE SOLUTION OR WAYS TO APPROACH THE ISSUE

Efficient policy design to address capital stock turnover requires an understanding of the investment decision. Policy measures that are efficient and effective will directly address the realities of the investment decision. Micro-scale analysis—the investment decision from the investor's perspective—is essential therefore for policy design. Such analysis can complement macro-scale modeling analysis.

While it is difficult to weigh the relative significance of the direct and indirect factors that determine turnover in models, it appears that most of the climate mitigation modeling analysis has used net present value criteria to choose the nature and timing of investment. This simplified criterion has the effect of overstating the rapidity of investment, as it fails to account for technological, economic, and policy uncertainty. Derivatively it may overstate mitigation, since it does not capture many of the considerations that go into an investment choice.

As the IEA (2007a and 2007b) has shown, if investors face uncertain policy development in addition to other market uncertainties and barriers, they may well choose to delay investment until the nature of the uncertainty is clearer (in other words, wait to acquire additional information). Uncertainty may also influence the nature of the investment decision in addition to technology choice. For example, if the time path of climate policy stringency is uncertain, investors may choose refurbishment over replacement. Refurbishment becomes a middle option, and the investment options therefore are less "lumpy and discrete" than it appears from models. While new capital may have performance characteristics closer to the state of the art, a refurbishment decision that makes a somewhat more modest improvement in plant efficiency may make more economic sense (Lampert et al. 2002).

In the IEA's Project Transitions, understanding the quantitative significance of the choice between refurbishment and new build is a central feature. This relationship is quite complex. Initial results (Blyth 2010) indicate that the value of refurbishing and waiting can be quite substantial. Among the preliminary findings:

- As older plants lose relative economic performance to new builds, they are most likely to initially move lower in the merit dispatch orders, rather than fully close down.
- Higher electricity and carbon prices will accentuate the relative attractiveness of new builds versus refurbishment.
- Reducing uncertainty about carbon prices, fuel and electricity prices increases the incentive for plant replacement.
- A high value of waiting may require stronger regulatory policies to encourage capital replacement.

Blyth finds several reasons in the literature for a bias toward delaying new builds even beyond the question of refurbishment options. These include the following (as reported in Blyth 2010).

The combination of uncertainty regarding investment decisions, the possibility of learning in the future which resolves some of this uncertainty, and some degree of irreversibility in the investment decision creates an option value for waiting, which encourages continued use of existing plant. (Dixit and Pindyck 1994)

The performance of new technology may be expected to improve over time (or risks reduce over time) due to technological spillovers of R&D and learning-by-doing by *rival firms, creating* an incentive not to be first-mover. (OECD 2000, emphasis added)

Companies may have a vested interest in particular technology types that are engrained in the industrial structure of the country in which they operate, which distorts their incentives to invest in best available technology. (Krusell and Ríos-Rull 1996)

Companies may benefit from maintaining diversity in the mix of technologies they use for generation, which together with other strategic objectives can override the economics of individual plant choice. (Mulder, de Groot, and Hofkes 2003)

DRIVERS FOR INVESTMENT DECISIONS

Considering the role of fuel prices relative to carbon prices, the IEA (2007a) found that, in most cases, fuel prices dominated carbon prices as determinants of the timing of the investment decision and of the technology choice. Blyth found that "reducing uncertainty with regards to carbon, fuel and electricity prices would tend to reduce the incentive to invest in existing plant, and would help to incentivise greater plant replacement rates" (Blyth 2010). A study on Germany (Schwarz 2005) found that failure to include modernization possibilities in models led to significantly different estimates of options.

A Pew Center study (Lampert et al. 2002) also concluded that other factors can be important drivers of the investment decision not easily captured in a net present value calculation. Firms were found to divide investments into mandatory and discretionary. Based on surveys of firms, that study concludes that discretionary investment decisions are determined by corporate strategic considerations that are determined by market conditions and corporate goals. Mandatory investments involved required maintenance and refurbishment, but also were driven by health, safety, and regulatory requirements.

In some cases, and in particular with electricity generating plants, the problem may be complicated by incompatibility of new technology with existing locations. For example, centralized solar energy facilities and wind farms are much more land-intensive than fossil-fuel fired plants. It is not possible for solar and wind technologies to replace existing fossil plants at the same location. New sites must be utilized. This complicates the decision to invest in these technologies as a replacement for existing fossil ones.

Finally, energy facilities, including renewable ones, frequently encounter delayed siting processes due to local opposition, among other factors. For example, a proposed off-shore wind farm in Nantucket ran afoul of local visibility concerns (see Daley 2006). Not-in-my-backyard concerns are a particularly powerful concern for investors (IEA 2007b). From the investor's perspective, therefore, the investment decision compares a repowering decision versus the superior greenhouse and possibly energy performance of new technology, which is encumbered by siting and regulatory delays. Again, this tends to tilt the investment decision in favor of refurbishment as existing sites become very valuable.

The investment decision is indeed more complex than is included by the net present value criterion. These factors and others work to delay the decision to invest and to discourage the introduction of the most carbon free technologies. Policy development needs to consider the investor's perspective in designing solutions to these barriers.

If the rate of investment is to accelerate and become more climate friendly, then policy will need to be developed in a timely fashion. But, in addition to the time required for the private sector to consider and develop a new investment, there is the lag in developing policy.

For example, the Kyoto Protocol was signed in 1997, whereas a dozen years later there existed only a negotiating text to cover the post-2012 period, which was to begin in only three years. Given the time lag required to implement an investment choice, this will significantly affect operating and investment decisions of firms. The realities of international and national policy formulation are that it is a messy, time-consuming process.

The principal reason for this delay is the strong differences among key emitters as to the path forward. Some of those differences may have been reduced as a result of the 2008 U.S. election, but other differences, namely between the industrialized, industrializing, and developing countries have yet to clarify the nature of a compromise. While negotiating text exists, the process forward is complicated by the existence of two negotiating texts in two different committees.[2] In mid-2009, the negotiating texts were fundamentally compilations of national submissions plus additions made at the June 2009 negotiating session in Bonn. In other words the negotiating text got longer, not shorter, even though less than six months remained before the November 2009 Copenhagen meeting.

It seemed evident, even before the Copenhagen meeting, that it would not produce the necessary detail to empower a mitigation future. Delegations were trying to find, in the current text or some alternative language, the key elements that would enable ministers to announce a success at Copenhagen. Even this problem was non-trivial, since what is essential to political success differs markedly among countries.[3] Whatever its outcome, subsequent negotiation elaborating the Copenhagen outcome seemed inevitable.

After sufficient international architecture exists to clarify legal obligations and to establish modalities under the new agreement, national legislation will be necessary to implement international obligations. There will certainly be differences between countries in this process, dependent upon political structure and previous legislative action. Nevertheless, the sectoral breadth of emissions and the likely supplemental regulatory structures that will accompany a price mechanism will require assembly of a national package of policies. And that will take time. It does not seem excessive to imagine legislative and regulatory rule-making taking several years if countries are to establish stringent new goals for reductions. This international and national policy development lag seriously compromises the markets' ability to deliver on the national commitments to reach the 450 ppm target. Of course, even when fully implemented, all policy measures will have lags before fully achieving their expected mitigation outcomes.

WHAT DOES THIS IMPLY FOR POLICY?

Countries sought a comprehensive agreement from the Copenhagen deal—a deal that would engender an effective global framework for mitigating and adapting to climate change and is affordable and internationally equitable. This chapter focuses only on the effective mitigation piece of this problem. Specifically, the focus has been on the difficulties in achieving timely mitigation through accelerated investment in lower carbon technologies both in greenfield plants and in refurbishments of existing infrastructure. It seems likely that policy packages will include incentives both for new builds and for refurbishments: for example, carbon prices and taxes, but also regulatory measures that apply either to new builds or to refurbishment. For example, building codes generally apply to new construction while other codes are needed to foster efficient refurbishment (such as codes for window replacements).

Beyond this general conclusion, IEA work, the international modeling work summarized by the IPCC in Figure 4.2, and the limited literature on capital stock turnover policies indicate that there is urgency for policy development if stated goals are to be met. In this sense, Copenhagen and its further elaboration are critical to preserving any option for emissions below 500 ppm.

There are several characteristics of a deal that are relevant for an effective agreement.

A Focus on Energy Efficiency

First, it must enable efforts to suppress energy demand while providing for growth in energy services: in other words, enabling energy efficiency policy is essential to any package. Restraining energy demand is indispensable for reducing emissions, and the growth in services is essential for maintaining public support for aggressive policy measures. Fortunately, the potential exists for substantial mitigation through energy efficiency policies. It will require regulatory policy measures to complement a carbon price, as has been documented in the literature.[4]

The potential for energy efficiency is amply demonstrated in the literature. For example, the IEA (Waide et al. 2006) has studied refurbishment in new high-rise buildings in Europe and found that a 28 percent

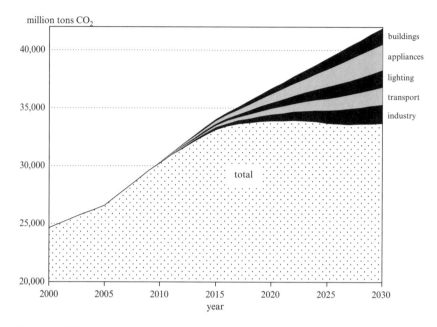

Source: IEA database.

Figure 4.2   IEA-25 energy efficiency recommendations, 2000–30

cost-effective energy savings potential exists. Price is important but not enough to exploit this potential. Market barriers exist, in particular the principal-agency problem (IEA 2007c) reduces the full range of options for responding to price changes. Other policies will have to be developed if the full potential is to be exploited. Certainly they include broader application of and more aggressive building codes.

In another study, the IEA looked more broadly than the European Union and estimated, using data from its *World Energy Outlook*, "that the total feasible potential for energy savings by renovation and refurbishment in most OECD countries will be around 50 percent of the actual consumption" (Laustsen 2008). In transition economies this potential will be even larger, because of the lower energy standard of the existing buildings. For developing countries outside the OECD, the feasible potentials are estimated to be larger too, but the savings in these countries will be reduced by an increase in the comfort levels for both cooling and heating.

Both IEA studies mentioned above found that renovation is more cost effective when timed to coincide with maintenance and refurbishment cycles in building.

The IEA has been assessing energy efficiency practice more generally than in the building sector alone in member and non-member countries. This work was carried out as a result of the G8 Summit in Gleneagles in 2005. Because of that mandate, the IEA has developed a set of 25 energy efficiency policy recommendations. These measures could have a substantial effect on emissions if implemented globally. Figure 4.2 shows the IEA estimate of the potential savings should the measures be immediately implemented globally. Of course, no policy will achieve all of its potential, but these estimates indicate that energy efficiency should be a priority because the cost-effective potential is so large. Energy efficiency policy has the advantage of "buying time" for capital with very long construction times or for the development of newer technologies with lower carbon emission profiles.

IEA analysis suggests that a carbon price is as critical for energy efficiency improvement as it is for inducing the structural changes required for a low carbon economy. However, market failures inhibit the attainment of the full, economically justified potential for energy efficiency. Principal-agent problems and information asymmetries are the most frequently identified market failures. The energy efficiency literature generally makes this case without providing evidence of the quantitative significance of market failures. The IEA has begun to document the potential size of some market failures. In *Mind the Gap: Quantifying Principal-Agent Problems in Energy Efficiency* (IEA 2007c), case studies identified the energy use in a particular end use (Table 4.1) that was potentially subject to principal-agent problems.

While some end uses are relatively unaffected, others (such as space heating, lighting, water heating, and refrigerators) are more significant.

Other end uses may also be subject to barriers and market failures. For example, consumer electronics and integrated communication technologies may also exhibit principal-agent problems and information asymmetries. Digital cable converters, sometimes referred to as set-top boxes, involve both cable providers and manufacturers in their specifications. Cable providers in particular and manufacturers to a lesser degree may not consider energy efficiency in their manufacture. Cable companies provide specifications for such boxes and have an interest in accessing the cable boxes of individual owners at any time throughout the day, so that they can download scheduling information to the television. They need the set-top box to be in standby mode for such downloads. It has not been unusual for the signal converters to use nearly as much energy in standby mode as in the active modes. The IEA (2009) estimates that 90 one-gigawatt power plants have been constructed worldwide to provide the standby power of our appliances.

Table 4.1   Principal-agent case studies

Use	Energy use affected by the PA problem (PJ/year)	Total sectoral energy use in 2005 in the relevant country (PJ/year)	Energy use affected by PA problem as a share of total sector energy use (%)
Residential sector			
Refrigerators, space heating, water heating, and lighting (United States)	3 546.5	11 296.5	31.4
House heating (Netherlands)	105.0	433.0	24.3
Set top boxes (United States)	68.4	11 296.5	0.6
Commercial sector: office space			
Japan	60.5	2 575.1	2.3
Netherlands	24.5	316.6	7.7
Norway	5.4	103.2	5.2
End-use appliances: vending machines			
Japan	6.1	2 575.1	0.2
Australia	1.5	243.5	0.6

Source: IEA (2007c).

Where market failures exist, there is a case for government intervention to correct the incentives. IEA experience has shown that (1) policies targeted at the specific market failures or barriers are most effective and (2) regulatory approaches, including information programs such as labelling, work most effectively when supported by prices. In this sense, a carbon price is also critical to energy efficiency.

In addition to micro-scale, end-use analysis of energy efficiency policy, the IEA has also considered the role of energy efficiency in macro-scale analysis. The *World Energy Outlook* (IEA 2008c) defines a reference scenario. Figure 4.3 presents the latest reference case.

Governments in the IEA region have taken mitigation policy measures over the last decade and a half. Policies that reduce emissions have also been taken in developed and developing countries. For example, China has implemented fuel efficiency standards for automobiles and has promoted the growth of renewable energy. The European Union has implemented an emissions trading regime in addition to regulatory directives to promote energy efficiency in buildings and promote renewables. Nevertheless, even

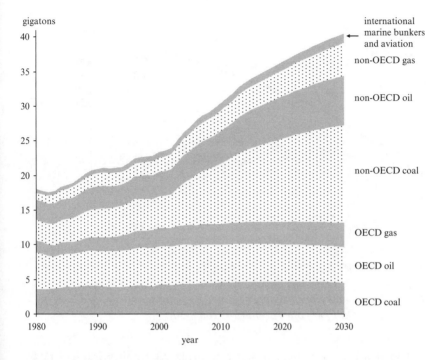

Source: IEA (2008c).

Figure 4.3 Energy related CO_2 emissions in the reference case, 1980–2030

with the policies, emissions have continued to rise and in fact there was an upward inflection just after 2000. Policies get implemented, and yet there is still no discernable influence on emissions trends. How can this be? Part of the answer can be seen in the regional source of emissions growth, which mirrors underlying changes in energy consumption.

When the IEA was founded in 1974, its member countries were not only the largest source of fossil-fuel demand but also a significant source of the increase in demand. This is no longer the case. The world energy order is changing as the emerging economies of China, India, Brazil, South Africa, Mexico, Korea, and Indonesia have become the marginal consumer. Ninety-seven percent of the increase in energy demand will be in non-OECD countries and nearly three-quarters will be in China, India, and the Middle East. Investment in a lower carbon future will be directed toward these countries in addition to the OECD countries. It is for this reason that some industrialized countries have taken strong negotiating positions with regard to the participation of these countries in the next control regime.

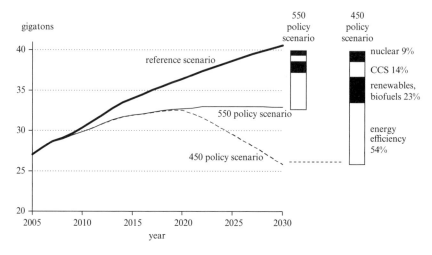

Source: IEA (2008c).

Figure 4.4 *Reductions in energy related emissions in the World Energy Outlook policy scenarios, 2005–30*

The *World Energy Outlook* has also explored a policy framework for meeting declared objectives for stabilizing concentrations: namely 450 and 550 ppm. The framework is built on the major mitigation building blocks identified in the Bali Action Plan (UNFCCC COP13 2007): namely, emissions trading, sectoral approaches, and national policies and measures.

As Figure 4.4 shows, energy efficiency provides the largest contribution to mitigation. This is not surprising, given the well documented monetary cost advantage of energy efficiency and given the more immediate effect of energy efficiency policy in many end uses. Supply side investments frequently take longer to complete than some of the end use efficiency.

Appliances and electronic equipment, some building refurbishments, and automobile fuel consumption savings can be achieved in a relatively shorter period of time when compared with the ten to fifteen years it takes to build a nuclear plant or even five years for a natural gas plant.

A Focus on the Main Emitters and Emitting Sources

Second, the geographic scope of emission sources is daunting and another complication for a new climate framework. All economic sectors contribute emissions, and therefore every country does as well. However, while climate change is frequently described as a global problem, and in terms

Table 4.2   Top 20 emitters from energy sources

Country	CO_2 (million tons)	Share of world total (%)
United States	5 696.77	20.3
People's Republic of China	5 606.54	20.0
Russia	1 587.18	5.7
India	1 249.74	4.5
Japan	1 212.70	4.3
Germany	823.46	2.9
Canada	538.82	1.9
United Kingdom	536.48	1.9
Korea	476.10	1.7
Italy	448.03	1.6
Islamic Republic of Iran	432.83	1.5
Mexico	416.26	1.5
Australia	394.45	1.4
France	377.49	1.3
South Africa	341.96	1.2
Saudi Arabia	340.03	1.2
Indonesia	334.64	1.2
Brazil	332.42	1.2
Spain	327.65	1.2
Ukraine	310.29	1.1
Total of top 20		77.8

Source: IEA (2008a).

of impacts it most certainly is, it is less so in terms of mitigation. Table 4.2 lists the 20 largest emitters of carbon and their shares of the global total. If an agreement had these emitters within a control regime, then nearly 80 percent of the global total would be included. Nearly all of the global energy technology development capacity resides in these countries as well, so those countries outside the major emitters would be using technology developed within this group. While including some newly industrializing countries is essential for environmental effectiveness, it is also the case that a focus on major emitters may be a simplification that contributes to a resolution to international negotiations and yet is consistent with environmental effectiveness. As time passes and emissions outside the top 20 rise, they could be added to the initial mitigation framework. The decision about the aspiration of the control regime may well be a decision taken primarily by these main emitters, thus simplifying negotiations to those that will incur the greatest cost from the control regime.

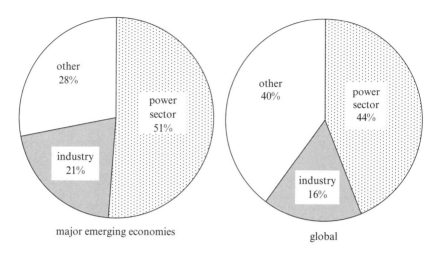

major emerging economies   global

Source: IEA (2008c).

Figure 4.5   Sectoral CO$_2$ emission share in 2010: major economies and the world

Of course, while most countries may not be included in the control regime (a national or sectoral emission cap), other incentives could be used to limit their emissions growth. Certainly, the clean development mechanism is one possibility currently in the international policy mix. However, bilateral or multilateral energy efficiency cooperation could play an important role as well.

An Intermediate Focus on Sectors

Third, it is possible that the sectoral coverage of an agreement may be initially limited. This limitation may not significantly reduce the effectiveness of an international agreement, particularly if such a configuration is directed at the critical sectors and is transitory in nature. Figure 4.5 demonstrates just how significant powerplant emissions are for the mitigation.

In work on sectoral approaches to mitigation, published as *Sectoral Approaches in Electricity Generation: Building Bridges to a Safe Climate* (IEA 2009), the IEA considers how a sectoral crediting approach could extend the breadth of participation in a control regime to emerging economies, thus increasing the effectiveness of an agreement configured like Kyoto.

This work finds that there is a considerable potential for avoiding carbon lock-in through the use of such a sectoral approach in the near term. Figure 4.6 shows just how large the lock-in could be. To 2030 it could

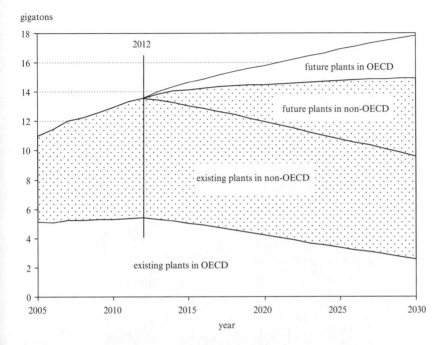

Source: IEA (2008c).

Figure 4.6 Projected emissions from electricity generation: OECD and non-OECD, 2005–30

be 4 gigatons. Given the significance of lock-in for the electricity generating sector, this approach may be a practical one for moving forward in the near term. While ultimately broader sectoral and international application of national cap and trade or taxation systems offers the most cost-effective emission reductions, a sectoral crediting system can start the process of applying such broader approaches in those sectors where data credibility is greatest and where emissions are largest.

A Focus on Technology Development and Diffusion

Finally, technology R&D and international diffusion of new technologies is essential. According to the *World Energy Outlook* (IEA 2008b), it is possible to achieve 550 ppm using existing technologies. In the 450 ppm scenario, however, technologies currently not available to the market at competitive costs will be required, in particular carbon capture and storage technology and electric vehicles. Research, development and

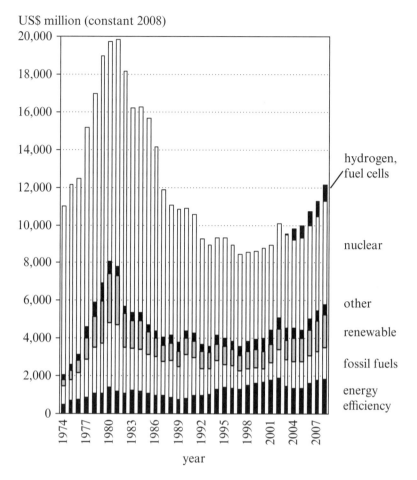

US$ million (constant 2008)

year

Source: IEA (2008).

*Figure 4.7   Public sector research, development, and demonstration
            expenditures for IEA countries, 1974–2008*

demonstration (RD&D) will be necessary to make these technologies ready
for the market in a regulatory, economic, and performance sense. Figure
4.7 shows public sector RD&D expenditures in IEA countries from 1974
to 2008. In response to the oil price shock of the 1970s, IEA governments
increased RD&D expenditures, in particular in nuclear energy, which is an
odd outcome. In the 1970s, a greater percentage of oil was used in power
generation than is the case today, but it is still surprising that emergency

conditions in the transport sector would increase R&D on electric generation technologies.

It is also interesting that there appears to be little increase in government RD&D expenditures even though it has been apparent that mitigation was essential since at least the early 1990s.[5] So, in spite of (1) the
transformation required in the energy sector, (2) considerable analysis
that the lowest numbers could not be achieved without the development
of new technologies, and (3) the declared priority of technology development by some IEA governments, there is little evidence of an RD&D
response. Reversing this trend is important to an effective climate and
energy policy response.

The IEA has been an advocate of increased technological cooperation for some time. It has over 40 implementing agreements under which
governments cooperate on RD&D. There are other RD&D cooperative
arrangements outside the IEA. Nevertheless, the economic scope of emissions and of the underlying capital structure that needs refurbishment or
replacement, or both, raises questions as to the sufficiency, effectiveness,
and coordination of these efforts.

The failure to significantly increase funding in the face of this unprecedented global challenge is a barrier to achieving the lowest stabilization
levels. Unless existing funding is used more efficiently than in the past,
there will be little hope for new technologies to contribute.

Effectiveness requires a more rapid diffusion of new technologies both
domestically and internationally than has occurred before. In particular,
carbon capture and storage is important for the continued viability of
coal. Of course, carbon and energy policies will have a dominant role to
play in changing the rate of capital turnover and technology diffusion.
The carbon price will be important, but in some cases, including carbon
capture and storage, government demonstration efforts can be critical and
potentially less demanding of national resources.

However, accelerating the RD&D process also depends on clarifying
early in the process the steps that are necessary for innovation and diffusion. To promote RD&D effectiveness by clarifying the requirements,
the IEA is developing 17 technology "roadmaps" that identify each of the
steps and the policies that can play a role.

New, more effective ways of cooperation can be helpful. While not dis-
paraging the important role competition plays in developing and diffusing
technologies, the unprecedented economic and geographic scope suggest
that no country can develop all the technologies that may be necessary.
Cooperation, at least in the early stages of R&D before preparing a technology for market competition, may economize on resources and also reduce
intellectual property concerns that may inhibit diffusion internationally.

CONCLUSIONS

Inertia in the capital stock is a significant obstacle to rapid changes in emissions rates. It is essential to develop policies and emissions management strategies that recognize these obstacles and seek to manage that capital stock transition with practical policies that target the lowest cost now, while also setting in place incentives to enable longer term changes.

Aggressive end use energy efficiency policies can potentially allow the alignment of refurbishment and replacement decisions with historic rates of capital turnover, thus lowering the cost of aggressive mitigation.

While a comprehensive agreement, founded on a price for carbon, was the preferred Copenhagen outcome in 2009, failure to achieve this outcome does not prevent aggressive pursuit of initial practical steps in the most important emitting sectors, in particular electricity generation.

Analysis in the IEA's *World Energy Outlook* finds that the most aggressive stabilization targets (such as 450 ppm) cannot be achieved without new technology development. Cooperation on at least the development and demonstration stages is essential, given the scope of new technologies that will be required.

Finally, the IEA 450 scenario shows a significantly different set of investments by 2030 compared with the 550 scenario (in particular, building gas-fired plants is not an option consistent with achieving 450). And since 2030 is not very far off, clear signals about direction are needed to help steer investment decisions accordingly.

ACKNOWLEDGMENTS

The author wishes to thank Richard Baron, Barbara Buchner, and William Blyth for providing comments, without ascribing any errors to them.

NOTES

1. See Blythe (2010:17, Figure 3) available in electronic form at www.iea.org/papers/2010/economics_of_transition.pdf.
2. One process is under the Kyoto Protocol and considers extensions and modification to that agreement. The second process is under the United Nations Framework Convention on Climate Change and considers elaboration in the form of new agreements to that document.
3. In developing a bottom-line for the November 2009 Copenhagen meeting, delegations were balancing politically necessary language for advertising success against concerns about prejudicing the outcome of the further negotiations that would enable the Copenhagen result.

4. See for example Laustsen (2008), Jaffe and Stavins (1994), Bailey and Ditty (2009), Worrell and Biermans (2005), and Levine et al. (2007).
5. It is difficult to say when governments were authoritatively warned that mitigation was essential, but certainly the IPCC first assessment report provided that advice and was the process that governments established to advise them on that question.

REFERENCES

Bailey, Ian, and Christopher Ditty. 2009. Energy Markets, Capital Inertia and Economic Instrument Impacts. *Climate Policy* 9 (1): 23–39.
Barker, T., I. Bashmakov, L. Bernstein, J.E. Bogner, P.R. Bosch, R. Dave, O.R. Davidson, B.S. Fisher, S. Gupta, K. Halsnæs, G.J. Heij, S. Kahn Ribeiro, S. Kobayashi, M.D. Levine, D.L. Martino, O. Masera, B. Metz, L.A. Meyer, G.-J. Nabuurs, A. Najam, N. Nakicenovic, H.-H. Rogner, J. Roy, J. Sathaye, R. Schock, P. Shukla, R.E.H. Sims, P. Smith, D.A. Tirpak, D. Urge-Vorsatz, and D. Zhou. 2007. "Technical Summary." In *Climate Change 2007 Mitigation. Contribution of Working Group III to the Fourth Assessment Report of the Intergovernmental Panel on Climate Change*, edited by B. Metz, O.R. Davidson, P.R. Bosch, R. Dave, and L.A. Meyer, pp. 25–94. Cambridge and New York: Cambridge University Press.
Blyth, William. 2010. *The Economics of Transition in the Power Sector*. IEA Information Paper. Paris: International Energy Agency. Available in electronic form at www.iea.org/papers/2010/economics_of_transition.pdf.
Daley, Beth. 2006. *Boston.com Local News* (3 April). Available at www.boston.com/news/local/articles/2006/04/03/cape_wind_project_faces_new_threat.
Dixit, Avinash K., and Robert S. Pindyck. 1994. *Investment under Uncertainty*. Princeton, New Jersey: Princeton University Press.
International Energy Agency (IEA). 2007a. *Climate Policy Uncertainty and Investment Risk*. Paris: International Energy Agency.
International Energy Agency. 2007b. *Tackling Investment Challenges in Power Generation*. Paris: International Energy Agency.
International Energy Agency. 2007c. *Mind the Gap: Quantifying Principal-Agency Problems in Energy Efficiency*. Paris: International Energy Agency.
International Energy Agency. 2008a. CO_2 *Emissions from Fuel Combustion*. Paris: International Energy Agency.
International Energy Agency. 2008b. *IEA Energy Technology RD&D Database*. Available at wds.iea.org/WDS/ReportFolders/ReportFolders.aspx.
International Energy Agency. 2008c. *World Energy Outlook 2008*. Paris: International Energy Agency.
International Energy Agency. 2009. *Sectoral Appoaches in Electricity: Building Bridges to a Safe Climate*. Paris: International Energy Agency.
Jaffe, Adam B., and Robert N. Stavins. 1994. The Energy Paradox and the Diffusion of Conservation Technology. *Resource and Energy Economics* 16 (2): 91–122.
Kahn Ribeiro, S., S. Kobayashi, M. Beuthe, J. Gasca, D. Greene, D.S. Lee, Y. Muromachi, P.J. Newton, S. Plotkin, D. Sperling, R. Wit, P.J. Zhou. "Transport and Its Infrastructure." In *Climate Change 2007 Mitigation. Contribution of Working Group III to the Fourth Assessment Report of the Intergovernmental Panel on Climate Change*, edited by B. Metz, O.R. Davidson,

P.R. Bosch, R. Dave, and L.A. Meyer, pp. 323–86. Cambridge and New York: Cambridge University Press.

Krusell, P., and J.-V. Ríos-Rull. 1996. Vested Interest in a Positive Theory of Stagnation and Growth. *Review of Economic Studies* 63:301–29.

Lampert, Robert J., Steven W. Popper, Susan A. Resetar, and Stuart L. Hart. 2002. *Capital Cycles and the Timing of Climate Change Policy.* Washington: Pew Center on Global Climate Change.

Laustsen, Jens. 2008. *Energy Efficiency Requirements in Building Codes: Energy Efficiency Policies for New Buildings.* IEA Information Paper. Paris: International Energy Agency. Available at www.iea.org/textbase/publications/index.asp.

Levine, Mark, D. Ürge-Vorsatz, K. Block, L. Geng, D. Harvey, S. Lang, G. Levermore, A. Mongameili Mehlwana, S. Mirasgedis, A. Novikova, J. Rilling, and H. Yoshino. 2007. "Residential and Commercial Buildings." In *Climate Change 2007 Mitigation. Contribution of Working Group III to the Fourth Assessment Report of the Intergovernmental Panel on Climate Change,* edited by B. Metz, O.R. Davidson, P.R. Bosch, R. Dave, and L.A. Meyer, pp. 387–446. Cambridge and New York: Cambridge University Press.

Mulder, P., H.L.F. de Groot, and M.W. Hofkes. 2003. Explaining Slow Diffusion of Energy-saving Technologies. *Resource and Energy Economics* 25:105–26.

Organisation for Economic Co-operation and Development (OECD) and International Energy Agency (IEA). 2000. *Experience Curves for Energy Technology Policy.* Paris: Organisation for Economic Co-operation and Development and International Energy Agency.

Pacific Northwest Laboratory. 2003. *Capital Stock Turnover Rates.* Richland, Washington: Pacific Northwest Laboratory.

Pacific Northwest Laboratory. 2004. *Fossil Fuel Resources.* Richland, Washington: Pacific Northwest Laboratory.

Schwarz, Hans-Gunter. 2005. Modernisation of Existing and New Construction of Power Plants in Germany: Results of an Optimisation Model. *Energy Economics* 27 (1): 113–37.

Stern, Nicolas. 2007. *The Economics of Climate Change: The Stern Review.* Cambridge: Cambridge University Press.

Stoneman, Paul. 2002. *The Economics of Technological Diffusion.* Oxford: Blackwell.

United Nations Framework Convention on Climate Change (UNFCC). 2007. UNFCCC COP13. FCCC/CP/2007/6/Add. 1 Bali Action Plan. Available at unfccc.int/resource/docs/2007/cop13/eng/06a01.pdf#page=3.

Waide, Paul, Pedro Guertler, and Winton Smith. 2006. *High-rise Refurbishment: The Energy Efficient Upgrade of Multi-story Residences in the European Union.* Paris: International Energy Agency.

Worrell, Ernst, and Gijs Biermans. 2005. Move Over! Stock Turnover, Retrofit and Industrial Energy Efficiency. *Energy Policy* 33 (7): 949–62.

5. The political economy of climate change

Lawrence Rothenberg

From a policy analytic perspective, in many respects the solution to climate change is straightforward: the producers of greenhouse gases must internalize the costs of their actions through implementation of an efficient policy instrument. Yet, atmospheric concentrations of greenhouse gases continue rising at alarming rates. It can be argued that political economic issues are a major component of the explanation for this contrast. Indeed, political economic issues constitute something close to a "perfect storm," where a multitude of problematic factors come together, so that producing an effective international solution to global warming is extraordinarily daunting. Jointly, the nature of the problem, the distribution of cost-bearers and beneficiaries among nation-states and generations, domestic policy decisionmaking incentives, and monitoring and enforcement issues present a formidable obstacle to dealing with global warming. Even given a post-Kyoto international agreement incorporating the United States and developing nation-states, there will be a host of questions about whether the agreement will be approved and implemented and whether it, and any subsequent agreements, will be sufficient. While global warming's potentially cataclysmic consequences have mobilized many voices and interests, the political economic pressures to do little must be overcome.

THE POLITICAL ECONOMY OF CLIMATE CHANGE

When scholars discuss the political economy of environmental policy, it is typically with respect to choices over policy instruments in advanced industrial economies—notably whether decisionmakers are inclined to choose relatively efficient market-based price or quantity instruments (and which type and with what specific provisions) or to adopt comparatively costly command and control methods (for example Keohane et al. 1999; Kirchgassner and Schneider 2003; Oates and Portney 2003; Stavins 2004a). Certainly, considerations about market versus command and control remain relevant, as illustrated by recent events in the United States and the use of command and control when market options exist—such as the raising of Corporate Average Fuel Economy (CAFE) standards. Nonetheless, market instruments will certainly be at the core of any substantial climate change mitigation.[1] Despite efforts by many to pursue their own self-interest without regard to social welfare, almost all parties appear to recognize that climate change mitigation is just too expensive without harnessing the efficiencies of the market to some degree. Thus, academic analyses of optimal instrument choice are typically over issues such as whether one wants taxes versus quantity markets (for example Weitzman 1974; Hepburn 2006; Mandell 2008) or some hybridized method (for example Murray et al. 2008) and whether grandfathering or auctioning permits is more desirable (for example Fischer and Fox 2007) or politically expedient (for example Lai 2008). Hence, focusing narrowly on instrument choice is a relevant, but not the central, political economy question for climate change.

Nevertheless, political economy is a prime consideration for understanding the climate change conundrum. But, rather than instrument choice exclusively, the political economic matters that are fundamental revolve around the desire and ability of nation-states to come to agreements and for subsequent implementation to be incentive-compatible, both at the present time and in the future.

As such, taking an admittedly broad and nontechnical approach, I overview key elements of the political economy of climate change. Specifically, my starting point is the contention that a key issue with respect to the political economy of climate change is how it impacts the potential for effective deals being struck among extremely diverse political actors given the nature of the global public good—climate quality—that is under consideration. By effective, I mean that the amount of mitigation chosen, even if it is not what policy analysts would select or does not adopt the precise instruments that they would favor, has two features: (1) it ameliorates the negative effects of climate change significantly and (2) the constructed agreement is executed in a manner so that policy goals are, indeed, realized now and in the future.

I begin with an analysis of the contrast between policy solutions to, and the reality of, climate change. I then elaborate considerably on the political economic ramifications that arise with trying to overcome the international collective action problem that climate mitigation, as a global public good, creates. Attention is then turned to implementation problems likely to arise once any climate agreement is struck in terms of its implementation. I conclude with some observations about what political economic approaches tell us about the likelihood of, and the obstacles to, effective climate mitigation.

POLICY SOLUTIONS VERSUS CLIMATE CHANGE REALITY

An initial point to recognize is that there is a tension between policy solutions for, and the realities of, global warming. At a very general level, policy solutions for climate change are straightforward. Conversely, the climate change situation is rapidly deteriorating, almost certainly reflecting problems associated with global collective action in tandem with technical, social, and economic constraints.

As for policy solutions, in most respects climate change is a textbook pure public goods problem: producers of greenhouse gases are not internalizing their actions' costs and, to a large degree, the benefits of a climate without anthropogenic degradation are nondivisible and nonexcludable (for example Stern 2008).[2] Whether or not persons, firms, institutions, or nation-states are producing greenhouse gases, or have produced them historically, is largely irrelevant for the costs that they are paying, or for the far greater costs that they will pay in the future, from a warming planet. Admittedly, from a static perspective, given the amount of greenhouse gases produced by a small number of political entities—notably China, the United States, and (if considered a single unit) the 27 members of the European Union (EU-27)[3]—a large member could somewhat reduce the costs that it would pay by substantially cutting back unilaterally or as part of a bargain struck among all the largest producers. However, as we will discuss, such actions would almost certainly be overwhelmed by future emissions by others without a global agreement.

Given this situation's textbook nature—and while ignoring a vast number of important issues—in its most general respect we have a straightforward economic solution to climate change. We want those producing greenhouse gases to internalize the costs of their actions, and we want them to do so by implementing an efficient policy instrument, be it a price (taxes) or a quantity (cap-and-trade) mechanism (for example Aldy et al. 2008). While the costs of mitigating climate change are high (although the numbers are frequently debated) and are shrouded by uncertainty associated with adaptation costs, if pricing is done properly there is reason to be optimistic about our ability to deal with this problem effectively. Innovation can be incentivized and social welfare can be maximized. For example, as many have done, we can calculate a cost for a metric ton of carbon (US$50 seems to be a standard estimate), incorporating discounting and equity considerations, and price accordingly.[4] Indeed, there are many, frequently citing the idea of "stabilization wedges" (Pacala and Socolow 2004; Socolow and Pacala 2006), who assert that the scientific, technical, and industrial know-how already exists and is waiting to be implemented (for example Gore 2006). That is, with proper incentives the problem is costly but solvable.[5]

Thus, for example, William Nordhaus (2009:3), a leading economic authority on global warming, states the policy solution succinctly:

> . . . we need to correct this market failure by ensuring that all people, everywhere, and for the indefinite future face a market price for the use of carbon that reflects the social costs of their activities. Economic participants—thousands of governments, millions of firms, billions of people, all making trillions of decisions each year—need to face realistic prices for the use of carbon if their decisions about consumption, investment, and innovation are to be appropriate.

Yet, such a confident tone flies in the face of casual observation that this solution is politically unfeasible and that the situation is quickly deteriorating. This is buttressed by more careful analyses, such as those conducted by the Intergovernmental Panel on Climate Change (IPCC), that find that climate change mitigation has been slow to move forward and that continuing accumulation of greenhouse gases in the atmosphere is exacerbating the situation. For example, there is a stark juxtaposition between Nordhaus's solution and the IPCC's assessment that: "There is high agreement and much evidence that with current climate change mitigation policies and related sustainable development practices, global GHG [greenhouse gas] emissions will continue to grow over the next few decades" (IPCC 2007:44).

Three observations, in particular, are notable. First, despite almost two decades of international efforts—not to mention initiatives at the nation-state and subnational levels and by businesses and nongovernmental organizations (NGOs)—atmospheric greenhouse gas concentrations keep rising quickly.[6] Atmospheric concentrations of carbon dioxide (CO_2) were at roughly 280 parts per million (ppm) in the pre-Industrial-Revolution era, grew slowly through the years until the mid-1960s, and, as Figure 5.1 illustrates, have taken a much steeper upward path subsequently and now stand at approximately 390 ppm.[7] This heightened growth has taken place against the backdrop of a growing elite consensus that anthropogenically produced climate problems exist (for example even American President Bush, a long-time skeptic, admitted this by 2007) and that there are potentially cataclysmic long-term consequences of not addressing them or doing so through ineffective, piecemeal actions (for example Oreskes 2004).

Second, the point at which policymakers propose that greenhouse gas levels stabilize seems to grow inexorably despite troubling scientific evidence suggesting that higher levels are even *more* problematic than earlier thought. As Figure 5.1 shows, CO_2 concentration has now raced past the

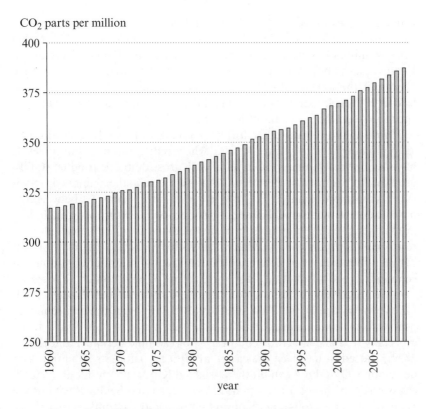

CO$_2$ parts per million

Note: The pre-industrial concentration was approximately 280 ppm.

Source: NOAA Earth System Research Laboratory data, National Oceanic and Atmospheric Administration, U.S. Department of Commerce.

Figure 5.1   Atmospheric carbon dioxide concentrations, 1960–2009

original 350 ppm level that many, such as James Hansen—the NASA scientist given principal credit for drawing attention to the relationship between greenhouse gases and global warming—originally focused on.[8] Despite warnings by Hansen and others that science suggests that going beyond 350 is more dangerous than originally thought (for example Hansen et al. 2008), the political focal point subsequently shifted to a 450 ppm stabilization point—for example for quite some time, this was the target of the European Union and many environmental NGOs (McKibben 2007). However, this goal has also largely (if not completely) gone by the board as well. Now 550 ppm has become the latest target *du jour*—the obvious reason being that it requires only about two-thirds of the emissions cuts of

450 ppm (for example Garnaut 2008). This failure to arrive at a firm global warming target is reflected by the less than optimistic statement by Nobel Prize winning physicist Steven Chu, U.S. President Obama's Secretary of Energy, that "I hope we hit 550 ppm" (Goodell 2009:61). Chu, a long-time believer in the 450 level, had to recast his vision, once he became a political leader and was looking at political realities, not to mention the vast differences between the price tags of lower and higher targets.

Third, unlike certain other forms of pollution (notably, conventional air pollutants such as sulfur dioxide), where reductions in pollution levels have traditionally gone hand in hand with gross domestic product (GDP) growth, the relationship between CO_2 production and economic development is more problematic. Put differently, there is a contrast between the observation that the greenhouse gas production keeps growing, including in many first-world countries, and the predictions of the so-called environmental Kuznets theory, which states that—akin to the relationship between inequality and economic development—after a critical point in economic growth is met, at least some forms of environmental degradation are effectively redressed—for example Grossman and Krueger 1995; for overviews, see Dinda 2004 and Stern 2004. (From a political economy perspective, it is also important to note that this has been found to be especially true in democratic countries—for example Bernauer and Koubi 2009.) Indeed, it has been the case that, at least before its definition as a pollutant, CO_2 did not exhibit the U-shaped Kuznets pattern but, instead, emissions kept rising (for example Holtz-Eaken and Seldin 1995; but see Schmalensee et al. 2004). Population numbers and economic activity have been the main determinants of greenhouse gas production, especially CO_2. Even if energy intensity per unit of GDP declines with growth (for example IPCC 2000), aggregate greenhouse gas emissions tend to rise with modest per capita income and population increases.

What explains these contrasts and the observation that, despite attempts to rein in greenhouse gas emissions and the assertion that it can be technologically accomplished with a multifaceted approach, the current trajectory for climate change is anything but promising? The argument of this chapter is that, when it comes to climate change mitigation, agreeing to implement a bundle of policies that can reasonably be claimed to be effective is a formidable task, and that political economy is a key part of the reason. The political economic issues associated with climate change constitute something close to a "perfect storm" where a multitude of problematic factors have come together to make coming up with a viable, enforceable set of agreements to deal with global warming extraordinarily daunting.[9] The very nature of the problem, the distribution of cost-bearers and beneficiaries among nation-states and generations, domestic policy decisionmaking incentives, and monitoring and enforcement difficulties, all present considerable obstacles to dealing with global warming. Even given a post-Kyoto international agreement that incorporates the United States and developing nation-states (the United States not being a Kyoto signatory and developing countries not being fully incorporated into the Kyoto framework), there will be a host of lingering questions about whether the political economic system will approve and implement that agreement, and whether that agreement, and those that will be subsequently necessary, will be sufficient.

CLIMATE CHANGE AS AN INTERNATIONAL COLLECTIVE ACTION PROBLEM

The starting point for understanding the political economy of climate change is to recognize that, besides being a global public good, it constitutes an international collective action problem. Successful mitigation involves nation-states working in tandem.

Put differently, the first rule of public economics as applied to the environment is that problems are best dealt with at the smallest political level capturing the vast preponderance of costs and benefits (for example Oates and Schwab 1988; Inman and Rubinfeld 1997). While this often leads to calls for a subnational delegation of authority (for example Anderson and Hill 1997), for climate change this logic suggests that a global political structure will be optimal. However, we are unlikely to move substantially to a purely global government at any point in the immediate future given that nation-states would be unwilling to abdicate sufficient amounts of their sovereignty.[10] Rather, we are, at most, likely to use some combination of treaties, which may or may not be binding in equilibrium, supplemented with supranational institutions that will condition their choices to what is accepted in equilibrium. Hence, the most profound political economic issue with respect to climate change is that it is a classic international collective action problem à la Olson (1965), although, as mentioned, China, the United States, and the EU-27 currently produce about half of all greenhouse gases. While the latter suggests some possibility of cooperation via bargaining, nonetheless some mechanism to get others to contribute to climate mitigation net of free-riding—the desire to let others pay while receiving full benefits—must be devised.[11]

As such, the starting point for any analysis of the political economy of climate change is to view it as a global collective action problem. Nation-states (or aggregates of them, such as the EU-27 or the Organization of Petroleum Exporting Countries, OPEC) must agree to voluntary contributions to the cause of climate change mitigation and then follow through on their pledges. While this is true of a variety of issues that are subject to international negotiations, it is the nuances with respect to climate mitigation that stand out as making dealing with global warming especially formidable. We turn to these specific features now.

Purity of the Public Good

> [G]lobal climate change is a public good [bad] *par excellence.*
> Kenneth Arrow (2007:3)

With some caveats, to be mentioned shortly, the good that we are talking about, climate, is truly a public good. Climate quality is nondivisible and nonexcludable. By contrast, while many other goods that are subjects of international collective action are seen as having many of the attributes of public goods, they actually can be made excludable or privatized to one extent or another.[12]

For example, defense alliances, such as the North Atlantic Treaty Organization—since defense is typically pointed to as the quintessential public good (Samuelson 1954; Olson and Zeckhauser 1966)—have been found to have excludable features, at least at specific points in time (notably when there is a movement away from using nuclear options to more modest means of defense).[13] The result has been that a more equitable distribution of contributions has been easier to generate because the threat of excludability is credible.[14] Alternatively, burden sharing for climate change is likely to fall heavily to large players with a high demand for mitigation because the threat of excludability is not credible. More importantly, the probability of provision of the public good is likely to be made more difficult because a smaller number of high demand nation-states cannot band together to exclude those who might be unwilling to contribute.

In a similar vein, in other international situations related to environmental quality (which are typically global commons issues in that goods are divisible), such as those concerning fish stocks and the quality of the world's oceans, it has been somewhat possible to take a Coasian (1960) approach by defining new, private, property rights. Most notably, by expanding the area off a coastline that could be claimed to belong to a specific nation-state—as an exclusive economic zone to 200 nautical miles, as was allowed with the 1982 Law of the Sea agreement—parts of the problem devolved directly to nation-states were more amenable to effective solutions. For example, a number of countries have then dealt with fish stocks using market-based quota systems with considerable success (for example Hannesson 2004; Newell et al. 2005). On the other hand, where environmental quality remains subject to global cooperation, large problems of declining water quality and crashing fish stocks continue to persist (for example United Nations 2009).

In short, the purity of climate as an international public good rules out solutions requiring excludability or privatization of divisible goods. Many of our seeming successes in inducing contributions from those who might have an incentive to free ride have seemingly been related to the ability to exclude or privatize. Without this, international bargaining becomes much more problematic.

The costs and benefits associated with climate change have at least three different, but interrelated, characteristics that have important political economic implications: (1) their sheer level, (2) their distribution across nation-states, and (3) their temporal incidence. We address each in turn.

Level of Costs and Benefits

Beyond its status as a public good, perhaps the most obvious feature of climate change is that there are huge immediate benefits to producing greenhouse gases and, potentially, very large immediate costs associated with mitigation. (I will defer discussing the benefits of mitigation for the time being, given they are principally in the future.) Whether or not investing in climate mitigation makes cost-benefit sense, or can be justified by something such as the precautionary principle that is particularly relied on in Western Europe,[15] politically it is well known that politicians are penalized for reducing the economic well-being of their citizenries (for example Lewis-Beck and Paldam 2000). The extent to which climate change mitigation will negatively impact the kinds of factors that citizens routinely judge politicians by—unemployment, growth, and inflation—can have profound political economic effects.

As already foreshadowed, everything being equal, the considerable costs of climate change mitigation will tend to push politicians, and even those lobbying them (Kolk and Pinkse 2007), to favor efficient instruments over more costly alternatives.[16] Thus, as mentioned, there has been relatively little dissent over the general principle of using cap-and-trade or taxes as a means to achieve climate change. Politicians often see cap-and-trade as desirable because it more effectively hides the true cost being incurred by their citizenry (hence, for example, opponents to climate change mitigation in the United States have tried to rebrand cap-and-trade as "cap and tax" to make the rough equivalence with taxation more transparent), particularly in nation-states without huge welfare sectors that require funding.[17]

Conversely, the linkage between how politicians are typically judged and the impact of actions to deal with climate mitigation will almost certainly severely diminish the willingness of politicians, *ceteris paribus*, to contribute to the global public good. Perhaps nowhere is this seen more starkly than in the United States. America would have paid the vast majority of the costs of climate mitigation, had it agreed to sign on to the Kyoto protocol—well over two trillion 1990 dollars, more than ten times the rest of the world's costs by Nordhaus's (2001) estimate—and, by virtue of its wealth and remaining by far the world's greater per capita producer of greenhouse gases net of OPEC nations,[18] is likely to be asked to pick up the largest tab for any new climate mitigation schemes.[19]

Thus, for example, when the U.S. House of Representatives passed legislation for diminution of greenhouse gas production below business as usual levels (but a decrease that is far less than that being suggested in world negotiations and would only put the United States back to 1990 levels by 2020), virtually the entire public debate was about costs. While proponents used claims that the costs were minimal (under $100 per family per year according to estimates from the U.S. Environmental Protection Agency), opponents employed far higher cost estimates (for example *Wall Street Journal* 2009). The degree to which politicians believed the low-end estimates, or thought that voters would punish them if such estimates were quickly discovered to be incorrect, was likely a key feature in determining the bill's ultimate passage.[20]

Although the contention that politicians will be hesitant to add substantial costs to their economy is an everything-else-equal argument, it might be countered that voters—particularly those in wealthier nations—also care about the environment, and that this negates the political costs associated with dealing with climate change.[21] As implied in our discussion of the environmental Kuznets curve, environmental quality is a normal good, in that citizens demand their government to produce more of it once an economic tipping point has been reached (Rothenberg 2002) and, presumably, would punish politicians not providing an appropriate level. However, as foreshadowed, there are several related caveats that make this argument less powerful with respect to global warming. For one thing, CO_2 emissions involve a global public good, and CO_2 has not had the U-shaped function associated with the Kuznets curve (although we obviously do not have a long history for the period after which there has been a scientific consensus regarding anthropogenic greenhouse warming). For another, consistent with CO_2 emissions' tendency to increase with economic activity, it is typically the case that those environmental goods that individuals are most willing to pay for have to do with features that they perceive as directly related to their well-being, such as clean air and clean drinking water. Conversely, the citizenry is much less enthusiastic about

Table 5.1   Americans' personal worries about environmental problems,
*   March 2009 (%)*

Problem	Degree of worry		
	Great deal	Fair amount	Only a little or not at all
Pollution of drinking water	59	25	16
Pollution of rivers, lakes, and reservoirs	52	31	17
Contamination of soil and water by toxic waste	52	28	19
Maintenance of the nation's supply of fresh water for household needs	49	31	19
Air pollution	45	31	24
Loss of tropical rain forests	42	26	32
Extinction of plant and animal species	37	28	34
Global warming	34	26	40

Note: The question asked was: "I'm going to read you a list of environmental problems. As I read each one, please tell me if you personally worry about this problem a great deal, a fair amount, only a little, or not at all. First, how much do you personally worry about . . .?"

Source: Gallup Poll, 5–8 March 2009.

paying for features that have a less immediate connection to their personal welfare (a category in which global warming certainly falls), regardless of whether the benefits of investing will exceed the costs.

This is reflected, for example, in American public opinion data where the citizenry are asked about how much they personally worry about a problem (Table 5.1). Strikingly, Americans concern themselves with drinking water and air pollution far more than global warming (or the decline of species or tropical rainforests for that matter), with only 34 percent worrying a "great deal" about global warming (Saad 2009; for a general overview of American public opinion on climate change, see Nisbet and Myers 2007).[22]

Interestingly, and in the same vein, cross-national data demonstrates that even citizen beliefs that climate change is anthropogenically driven are not a given, even among those who claim to know something about the issue—rather, it appears to be a positive function of the use of energy per GDP dollar in the economy (Pelham 2009b; for related findings about public concern regarding global warming, see Sandvik 2008).[23] Thus, for instance, the fact that only 49 percent of Americans who claim to know something about global warming believe that human actions are responsible is almost perfectly predicted by the intermediate energy intensity of its economy. There are a few outlier countries, most notably South Korea, which has low energy intensity but where 92 percent of the extremely knowledgeable population (93 percent of those surveyed) believe that global warming is a function of human activities.[24]

Intriguingly, then, those believing that global warming is a function of human activities are from nation-states that would likely have to cut the energy intensity of their economic actions (controlling for GDP) less than the norm. This, perhaps nonintuitive, relationship between the production of greenhouse output and citizen denial may be a function of both psychological processes—the desire to not acknowledge uncomfortable truths (Sandvik 2008)—and the strategic choices made by elites trying to maintain a favorable standing with the citizenry. In other words, at least some politicians, knowing the ramifications of the financial choices that would be involved by a citizenry clamoring to abate global warming over the past decade, may try to persuade their constituents that they need not be responsible for greenhouse gases. Thus, for example, as Republican leaders in the United States have spent more time discussing global warming, fewer Republican partisans, especially educated ones more likely to be aware of what their leaders are saying, accept the idea that global warming is a function of human activity (Pew Research Center 2008). This is consistent with the advice that Republican advisor Frank Luntz gave to party leaders in a now infamous 2000 memorandum that was obtained by the Environmental Working Group (see www.ewg.org/node/8684), an American NGO, where Luntz told party leaders to doubt global warming based on its scientific foundation.[25] This split of opinion is also in accordance with what we know about the generation of public opinion generally (Zaller 1992).

Thus, in terms of political economy, high immediate costs work against doing anything about climate mitigation in ways not captured by economic accounting. While substantial costs may facilitate adoption of efficient instruments (and make inefficient choices that much more problematic), political economic factors make political leaders hesitant to act sufficiently, given the metrics on which they are evaluated. Furthermore, the fact that the benefits of global warming are far more abstract than, for example, pure drinking water, detracts from citizens being willing to balance the economic costs imposed with the benefits of climate mitigation. Additionally, while perhaps endogenous to political debate itself, the fact that those from wealthy, carbon-intense economies question the need to act provides a further obstacle to effective collective action.

Distribution of Benefits and Costs across Nation-states

Related to the level of benefits and costs of global warming on the one hand and to mitigation on the other, another key issue with respect to solving

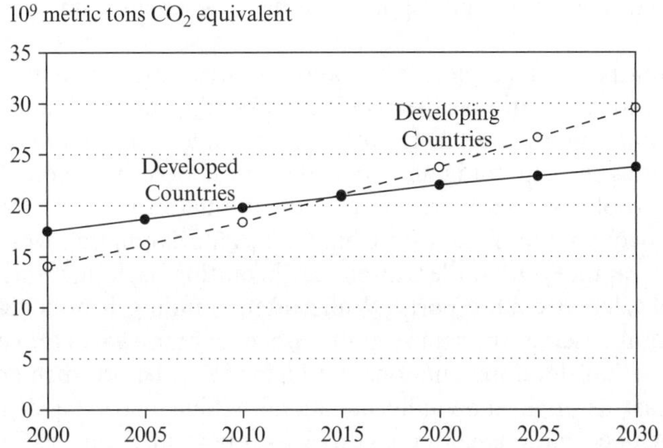

10⁹ metric tons CO_2 equivalent

Source: U.S. Environmental Protection Agency (www.epa.gov/climatechange/emissions/images/total_emissions_region.gif).

Figure 5.2 *Total greenhouse gas emissions by developed and by developing countries, 2000–30*

the international collective action problem involves the distribution of these benefits and costs across nation-states. Three features are particularly notable. First, a small number of players (especially if the EU-27 is considered to be one entity) currently produce most of the world's greenhouse gases. Second, just as China has supplanted the United States as the world's greatest producer of greenhouse gases, while currently developed countries produce the majority of greenhouse gases, with economic and population growth being drivers of greenhouse gas outputs, developing nations will exceed their total within the next decade given their higher economic and population growth rates (Figure 5.2). Finally, the relative costs of not mitigating greenhouse gases will fall especially on these same developing nations (see Ball 2008 for one set of estimates), as their economies are less well situated to adapt to changing temperatures. As stated in the United Nations Framework Convention on Climate Change (UNFCC 2007:5):

> Developing countries are the most vulnerable to climate change impacts because they have fewer resources to adapt: socially, technologically, and financially. Climate change is anticipated to have far reaching effects on the sustainable development of developing countries including their ability to attain the United Nations Development Millennium Goals by 2015.

While the first feature might provide reason to suggest that the global public good is amenable to a bargained solution, the latter two generate substantial obstacles that appear to have a greater impact on net emissions.

The first point is simple enough and requires little explanation. If we were to adopt a static perspective, the possibility of the United States, China, and the European Union, for example, bargaining with one another might help mitigate the collective action problem. The one obvious problem is that there are substantial differences in climate quality demand between the three, with China, a totalitarian nation with a much lower per capita GDP, far less interested in cutting back than their democratic, developed counterparts. While often standing behind the moral claim that it deserves the same right to use greenhouse gases for economic prosperity that developed nations did historically, before such emissions were recast as environmentally damaging, China's strident demands— that developed nations cut emissions by 40 percent from the 1990 level in roughly a decade while simultaneously handing over 1 percent of their GNP annually—are consistent with the posturing of a nation that can credibly be seen as willing to tolerate a level of prospective climate change that high demand countries are not.[26]

The second point, involving who the future producers of greenhouse gases will be, is only somewhat more nuanced. It is obvious that any deal by the major developed nations can be undercut by the actions of developing countries. As such, these less economically developed nation-states must inevitably be brought into any deal agreed to by developed countries. However, as these are low demand nations, the cost of getting them to go along promises to be high and requires either tempting incentives (for example cash payments, technological help, the ability to sell rights to pollute to developed nations, and even long-term obligation deferral until some income baseline is met) or serious threats (most obviously, trade barriers) that are, nonetheless, enforceable in equilibrium.

The final point, that developing countries will pay the biggest price relative to the size of their economies net of effective mitigation, while perhaps giving the developing states more incentive to come to agreement if their citizens and leaders do not discount the future too much, should also disincentivize those in the first world who must pay and reduce their standards of living. Developed nations, with far more ability to adapt their economies and societies to changes brought about by global warming, will suffer a smaller decline in their standards of living than their third world counterparts. This creates a difficult tension with respect to the bargaining situation created in trying to deal with climate change—those with the greatest costs not only have the least ability to pay but the least willingness to pay, *ceteris paribus.*

Therefore, from the standpoint of collective action, developed nations typically thought of as high-demanders for environmental quality have two reasons to not take onerous burdens upon themselves for climate mitigation: without buy-in from developed nations, their efforts will be overwhelmed by the actions of the developing world, and the benefits that they will receive from mitigation, while substantial, will be less than might be assumed. The latter, in turn, will create an even greater hurdle to overcome than might otherwise exist, for political leaders trying to convince their supporters to subsidize the bad behavior of those in developing nations.

Temporal Incidence: Time Consistency and Credibility

From a strictly policy analytic point of view, the major issue with regard to assessing temporal costs and benefits is what the right discount rate is to choose (for example Newell and Pizer 2003). As is well known, there is a vigorous debate about the correct rate, with a huge range of alternatives offered up as appropriate, each of which has important implications for what policy ought to be adopted (for example Sunstein and Rowell 2007).

However, from a political economic perspective, the key consideration that goes with the temporal incidence of costs and benefits revolves around time consistency and credibility (see, for example Aldy et al. 2001; Hovi et al. 2009). The basic problem is simple but has profound implications: Decisionmakers may elect to make comparatively modest changes now linked to the promise of more dramatic future commitments. When the time comes to follow through on these more dramatic changes, however, the successors to these initial decisionmakers may find them unpalatable. Minimally, they require an extreme technological optimism assumption, as they need innovation to drive the price of carbon mitigation down dramatically. If this is not realized, the initial decision to ramp up efforts later will likely not be credible.

As is probably obvious, the desire to take rather modest initial steps with the promise of more dramatic efforts later is extremely tempting for reasons of political expediency (or pragmatism to use a less pejorative term). In developed countries, domestic constituencies are sensitive to the costs that they must pay to benefit largely future generations and believe that, without buy-in from developing nations, their efforts may be marginal as well. In developing countries, resources are scarce and concerns about climate quality are weak. So an obvious path is to ask the developed countries to make somewhat modest efforts initially, with the promise of greater commitments later, and the developing nations to do little that is initially costly but much more when they reach development thresholds.

This, for example, has essentially been the proposal of Robert N. Stavins, the Harvard scholar who is perhaps the leading environmental economist in the United States (Stavins 2004a and 2004b; Olmstead and Stavins 2006; for a similar approach, see Nordhaus 2007). Stavins has advocated a "three-part global policy architecture." Recognizing that nations will have an incentive to free ride partially or fully, he argues for moderate but firm short-term targets and more ambitious but more flexible longer-term goals. He asserts that, if this is done, the credibility and time inconsistency problem can nevertheless be overcome.

> Can countries credibly commit to the long-term program that is part of this proposed architecture? Our answer is that, once nations have ratified the agreement, implementing legislation within respective nations would translate the agreed-upon long-term targets into domestic policy commitments. And once such implementing legislation was enacted, signals would be sent to private industry, which would begin to take action. Ultimately, such domestic actions provide the signals that other countries need to see. This represents a logical and ultimately feasible chain of credible commitment. (Olmstead and Stavins 2006:36)

The problem with this logic is that it rests on three potentially very problematic suppositions. The first is that the domestic choices to implement short-term goals, even if made, will be interpreted by investors as sufficient to warrant long-term technical and capital investments to meet the far more stringent rules to come (even though these rules must be kept flexible because of the large uncertainties of projecting into the future). The second is that the investments made will be sufficient to lower the costs of mitigation dramatically, thus making the much stricter and more difficult future reductions economically palatable. The third is that developing countries are committed to joining their first-world counterparts when their agreed-to development level is reached (or developed nations will then enforce sanctions to make them). This commitment presumably would exist even in nations suffering from political instability, be it from coups, transitions from one form of government (military, religious dictatorship, democracy) to another, or other forms of leadership change.

Alternatively, it would be easy for investors to believe that hard choices are being left to the future because there is a lack of political resolve. Similarly, claims by developing countries that they will initiate programs mimicking those of the developed countries in the future, when they reach certain benchmarks, may be interpreted as nothing more than cheap talk designed to pacify those in the first world and, even if sincere, may not be viewed as binding by successors in a manner comparable to how treaty obligations are often accepted in developed nations. Economic actors, recognizing the incentives of leaders to leave grandiose obligations for their successors who will lack incentive to implement them, may invest insufficiently, undermining the likelihood that first-world nations will sanction their third-world counterparts and creating the dreaded time inconsistency problem.

AFTER THE AGREEMENT: OBSTACLES TO SUCCESSFUL IMPLEMENTATION

As the discussion of credibility and time consistency implies, even should an agreement get hammered out that provides the hope of stabilizing greenhouse gas emissions at 550 ppm or less, a variety of political economic issues—notably domestic politics and enforcement and compliance—pose potential obstacles to successful implementation. While these related considerations loom large, our discussion will be somewhat brief and speculative, given a lack of specifics about the agreement in question.

Domestic Politics

One set of considerations revolves around domestic politics and the so-called two-level games (Putnam 1988) that incorporate international and domestic bargaining. While we have already surveyed a number of domestic political considerations, at least two others deserve elaboration: the tendency to choose inefficient policy instruments, which may increase the costs of climate mitigation and not create the kinds of incentives that will be needed for innovation, and the proclivity to pay off societal interests, who stand to lose when the costs of their actions are internalized, via sidepayments.

As mentioned in the introduction, choices of policy instruments have been a core political economy concern for quite some time. While less central for the political economy of climate mitigation, they are nonetheless relevant. Without a long discussion of why inefficient instruments are often chosen over efficient choices, it is enough to recognize that command and control still has an allure to politicians under certain circumstances and that the necessary conditions may, indeed, exist if climate mitigation responsibilities are delegated.

For example, the United States saw an example of poor instrument choice in May 2009 when President Barack Obama announced plans to hike the Corporate Average Fuel Economy standard (CAFE, a weighted average fuel economy for vehicles) from the previously mandated 27.3 miles per gallon in 2011 to 35.5 miles per gallon by 2016. Moreover, for the first time, the CAFE program was directly linked to a desire to reduce U.S. greenhouse gas emissions (the new standards are expected to reduce CO_2

emissions by 900 million metric tons from business as usual), the increase had widespread support, with auto manufacturers (seeking harmonized standards for the entire nation), organized labor, numerous state government officials, and environmental groups all in favor. Obama proclaimed the action "an historic agreement to help America break its dependence on oil, reduce harmful pollution, and begin the transition to a clean energy economy." What is normatively troubling is that CAFE is a command and control mechanism, and scholars have long shown that it is inefficient. One analysis estimates that the cost of reducing gasoline consumption by taxation is about one-sixth the price of using CAFE standards (Jacobsen 2008). In terms of innovation, the obvious problem—in contrast to a carbon tax, for example—is that, once a manufacturer meets the standard, its incentive to go further is greatly reduced relative to when one pays per unit of emission. Yet, CAFE standards appear politically attractive because they hide the implicit tax that goes with them (for example a 2006 Pew Poll found that 86 percent of the public favored raising the standard).

Analogous, in that it creates policy inefficiencies, is the ability of societal interests (for example agriculture in developed countries; see Lohmann 2003), often mobilized organizationally in rent-seeking activities, to garner sidepayments even when seemingly efficient instruments are selected. Again returning to the United States for illustration (although the European Union's implementation of the Kyoto protocol yields many similar examples), in the recent Waxman-Markey bill, concessions were made to coal-burning regions, the automobile industry, and agricultural interests (to name just a few) to move the cap-and-trade proposal forward. (For a particularly scathing review, see Spruiell and Williamson 2009.)

Whether domestic politics results in the adoption of inefficient instruments or the addition of sidepayments to make efficient instruments more palatable, the effect is similar. Domestic political considerations will raise the costs of climate mitigation compared with the efficient alternative. This will make climate mitigation more expensive (estimates of the costs of mitigation typically assume efficient instruments are chosen), and the likely need to use the wrong instruments or pay off rent-seekers will reduce political leaders' willingness to mitigate, everything else being equal, because they are judged principally on their economic performance and only secondarily on the environment.

Domestic political considerations may also undermine the desirability of giving nations or alliances of nations flexibility in implementation of some agreed-to reduction in greenhouse gases—rather than, for example, being committed to an internationalized harmonized cap-and-trade system or a tax. The more flexibility there is, and the less binding international commitments are, the greater the possibility that domestic choices will raise mitigation costs and decrease the resolve of political leaders. This may reinforce the time consistency and credibility problem discussed earlier. While, in the short term, allowing future flexibility may make it easier to move forward, in the longer term it can produce resistance by raising costs, particularly if the agreement is for modest initial steps followed by more dramatic subsequent commitments.[27]

Enforcement and Compliance

The fundamental problem of compliance in world politics is that it is virtually impossible to enforce international rules against powerful states. This in turn generates a lack of credibility of such rules *ex ante* (Keohane and Raustiala 2008:7).

Related to concerns associated with domestic politics are issues of enforcement and compliance. As Keohane and Raustiala imply, credibility of agreed-upon rules is a major issue in trying to solve global public goods problems like climate change. Even among the E.U. nations implementing the Kyoto accord, a variety of compliance issues have arisen, although, consistent with Keohane and Raustiala, the countries not meeting their obligations have felt little pain as a result (for example Barrett 2009).

But whatever complications exist regarding enforcement and compliance among E.U. nations, will almost certainly be dwarfed by those involving a global solution demanding participation from virtually all nation-states (for example Evans and Steven 2009). As already elaborated, not only will nations have free riding incentives, but they will have considerable diversity in their demand and willingness to pay for climate quality. Beyond that, regardless of whatever assistance is given by the developed nations, the capacities of developing nations will almost certainly be lacking in various degrees, meaning that these nations will not necessarily have the needed abilities (even should they have the will) to monitor or enforce complicated climate change commitments by themselves.[28] Indeed, many such nations lack even the basic capacity to define and enforce property rights that are basic to any international solution working smoothly.[29] In turn, this suggests that monitoring and credible sanctioning of noncomplying nation-states will be necessary for an effective climate mitigation regime.

Enforcement and compliance considerations have implications for the kinds of policies that are likely to work, and what one cannot help noticing is that they are at odds with what is usually associated with political pragmatism. Most obviously, the simpler and more transparent the policy chosen, such as an integrated carbon tax, the easier the monitoring and compliance tasks will be. Alternatives, such as a clean development mechanism, will likely be far less transparent and will be ripe for non-compliance and corruption (for example Michaelowa and Jotzo 2005).[30] Speculation has been raised that even using cap-and-trade will be too ripe for manipulation in developing nations, for example inducing governments to allow far more credits than the cap. Yet, as mentioned, more complicated alternatives are what are advocated in the name of producing palatable global policy architecture.

Regardless of what is chosen, assuming that there is an ability to monitor, there must be a willingness of members in the international community to entice compliance and enforce obligations even when it is not in their short-term interests. To generate compliant behavior, at least in the developing world, the most obvious solution is wealth transfers of the sort that China has been advocating. Presumably, such monies could be withdrawn in the face of noncompliance. The problem with such a mechanism is obvious: it can undermine the willingness of developed nations to agree on a mitigation scheme, because the policy will be both less attractive in a cost-benefit sense, and far more difficult to sustain politically given a hostile public. Another obvious solution is to utilize trade barriers; for example, if the United States does not sign on to a new global agreement or does not comply, an energy tariff could be placed on American exports (for example Stiglitz 2006). However, despite success with the World Trade Organization and the General Agreement on Tariffs and Trade before it, there is considerable reason to doubt the credibility of trade sanctions, given the high costs that would be imposed on consumers in participating countries and the potential for retaliation (Duval 2008).

Thus, assuming that a global mitigation agreement is agreed to, enforcement and compliance are likely to be a significant obstacle. Part of the reason is generic to international accords, while part is specific to the situation involving climate change. The kind of agreement that is likely to be ratified is one where monitoring is difficult, making noncompliance attractive. The fact that so much of any success will hinge on developing nations' promises to make successful enforcement even more challenging. And the high costs of any rewards or sanctions will make it difficult for compliance to be sustained. While actions to enforce compliance will likely take place, how effective they will be remains to be seen.

CONCLUDING THOUGHTS

Broadly, political economy helps us explain why some policies succeed and others fail (for example Glazer and Rothenberg 2001). Viewing climate change through a political economic lens suggests that political economic elements will tend to undermine the potential for the coordinated global sacrifices that are required for climate mitigation.

Put differently, from a political economic point of view, a variety of important elements will contribute to the potential for climate mitigation to be successful, for example for atmospheric concentrations to stabilize at 550 ppm or less. The citizenry must be convinced that anthropogenic climate change is real and that climate mitigation is of considerable personal importance; otherwise, they are likely to penalize political leaders when they implement costly policies that internalize costs. Political leaders will need to fight for efficient policies and policy instruments, so that the costs to the citizenry are kept down. There must not only be near universal participation in a climate change agreement, but the commitments made must be viewed as credible by decisionmakers so that investment capital and innovative effort are allocated and time inconsistency is not a looming problem. Developed and developing nations must be made to do as much up-front in the name of climate mitigation as possible, so that credibility and time consistency problems will not undermine international efforts. Monitoring mechanisms must be transparent and easily utilized so that credibility is buttressed. Nation-states must find it incentive-compatible to enforce agreements for credibility to be high and mitigation to have a strong chance of success.

Many of these suppositions contrast with what we know as social scientists and with what we observe in the real world. Many citizens have not yet been convinced that climate change is of great personal importance, despite a large accumulation of findings that global warming is real and is dangerous to the welfare of humankind. Political leaders have incentives to (and do) succumb to building inefficiencies into their policy choices. While a post-Kyoto agreement may or may not have high participation rates, any accord will likely have significant credibility and time consistency problems. And monitoring, compliance, and enforcement issues are likely to be major stumbling blocks.

While one can hope that climate mitigation can be a reality without decreasing the standard of living of the first world substantially and allowing developing countries to continue along a trajectory of economic growth, the obstacles are formidable. We should know even more clearly in the next few years in which direction we appear to be headed.

NOTES

1. For example, the Kyoto protocol involved a cap-and-trade system which imposed national caps on the emissions of so-called Annex I countries (developed nations

ratifying the agreement). Similarly, unsuccessful U.S. proposals voted upon in Congress (the so-called McCain-Lieberman proposals) also relied on markets, as did the 2009 Waxman-Markey legislation (passed by the House of Representatives in 2009 but not by the Senate). Virtually all discussions of a Kyoto successor treaty, for negotiations in Copenhagen in December 2009, had cap-and-trade or taxes as a core ingredient.

2. As I will discuss shortly, one additional feature is that the costs of climate change are asymmetrically felt by developing nations.

3. See the graphic illustration by the World Resources Institute (cait.wri.org/figures/ ntn/2-3-thumb.gif) of 2005 global GHG emissions, showing that the three largest and fifteen largest emitters account, respectively, for nearly half and nearly 80 percent of all emissions, whereas the rest of the world (for a total of 188 countries) accounts for the remainder.

4. For a meta-analysis of the various efforts to calculate such a social cost, see Tol (2008). It is interesting to note that an organization such as terrapass (which solicits voluntary contributions to offset one's greenhouse gas production) currently claims that it can offset 1,000 pounds of carbon for US$5.95 (see www.terrapass.com/faq/carbon-offsets).

5. Pacala and Socolow lay out seven different wedges, which collectively could contribute to a stabilization of CO_2 levels at 500 ppm over a 50-year period, which represents less than 50 percent of what emissions would be under "business as usual" without mitigation efforts, although it would still induce a substantial amount of climate change according to most scientific models.

6. International efforts to deal with climate change were initiated at the 1992 Earth Summit in Rio de Janeiro, where the United Nations Framework Convention on Climate Change was signed. The Kyoto protocol was agreed to in 1997, but did not go into effect until 2005, and expires in 2012. It was followed by additional national efforts (for example U.S. efforts to reduce its economy's energy intensity), subnational initiatives (for example the Regional Greenhouse Gas Initiative of northeastern and mid-Atlantic American states; see, for example www.rggi.org/home), and corporate responses (for example the well-known rebranding of British Petroleum to "Beyond Petroleum") spurred by the efforts of NGOs in what Baron (2001) has labeled "private politics."

7. CO_2 is the most prominent greenhouse gas, but other such emissions include methane, nitrous oxide, sulfur hexafluoride, perfluorocarbons, and hydrofluorocarbons. Including non-CO_2 gases typically increases the emissions shares from developing countries (Baumert et al. 2005).

8. Reflecting the frustration of watching efforts to focus on 350 ppm wane, on 23 June 2008 (the twentieth anniversary of Hansen telling the U.S. Congress about global warming), Hansen was among the signatories to full-page advertisements published simultaneously in the *Financial Times*, the *International Herald Tribune*, the *New York Times*, and the Swedish paper *Dalarnas Tidningar*, starkly calling for "350" to be the goal for CO_2 atmospheric concentration. While the advertisement's message was directed toward nations involved in the negotiations leading up to and beyond the Copenhagen Climate Change Conference in November 2009, it seems to have had little impact.

9. The term "perfect storm" typically refers to the simultaneous occurrence of weather events which, when they occur simultaneously, produce a violent and dangerous storm.

10. On the desirability of global environmental governance and mechanisms, see for example Esty and Ivanova (2002).

11. China's status as a developing nation and the previously mentioned possibility that increased GHG production by other nation-states could outstrip cutbacks by these major producers provide additional difficulties for coming to a collective bargained agreement spearheaded by large producers. As will be discussed, should the alternative that China and other developing countries are allowed to ramp up their mitigation efforts very slowly be chosen as a bargaining solution, there will likely be significant time consistency problems in the future.

12. The one good with a similar purity to that of global warming, and which scholars often like to contrast with the case of global warming, because it is an instance where international cooperation seemingly worked, has to do with the decline of the upper atmosphere ozone layer (for example Barrett 2007; Sunstein 2007). However, there are several meaningful differences between the two. First, only a very few industrialized nations were producing chlorofluorocarbons, the substance used in refrigerants and aerosols that was dissipating the upper-atmosphere ozone layer. Second, to foreshadow our analysis of demand for global warming mitigation, the citizenry of these nations believed that their personal well-being was at stake due to the harm that would be caused by increased exposure to ultraviolet rays. Third, there was a ready, and relatively inexpensive, substitute that had been developed by Dupont. None of these characteristics—limited number of producers, high levels of citizen concern regarding their personal well-being, and the ready availability of a financially manageable substitute—applies to the situation involving GHG mitigation.

13. Of course, defense alliances also typically cover only a subset of nations, often with very similar demand levels for the good in question, making coming to an agreement easier.

14. For example, Sandler and Hartley document how some forms of strategic doctrine in use at certain points in time allow exclusion more readily, producing more equitable burden sharing arrangements. They conclude that "the greater is the ratio of excludable benefits in terms of overall benefits within an international collective, the less suboptimal will be the outcome of members' independent behavior" (Sandler and Hartley 2001:893).

15. The precautionary principle essentially stipulates that one should weigh especially heavily the possibility that an action could cause irreversible harm to the environment. For a discussion, see Sunstein (2005).

16. Of course, everything is not always equal, and other domestic political considerations frequently point politicians to favor inefficient alternatives, creating yet another problem for effective climate mitigation by raising its cost. Additionally, even should seemingly efficient instruments be chosen, they may be subject to rent seeking behavior, such as efforts to get political leaders to distribute excessive numbers of emissions rights in a cap-and-trade system (for example Nordhaus 2007).

17. One potential problem with cap-and-trade relative to carbon taxes is ease of enforcement, particularly in less developed nations.

18. See for example the World Resources Institute estimate of CO_2 emissions per capita in selected economies in 2030, compared with 2005 (www.wri.org/image/view/9255/_original).

19. As of this writing, for example, China is demanding roughly ten times the GHG reductions by 2020 that the U.S. Congress is considering and an additional annual cash transfer of 1 percent of GDP (roughly US$140 billion currently) to developing countries.

20. Another reason why this legislation succeeded was likely the bargaining situation created by the American courts. The U.S. Environmental Protection Agency (EPA) was told by the U.S. Supreme Court in 2007 (in *Massachusetts v. EPA*) that, net of legislation, it had to either regulate GHG or deny the science of climate change. This gave the agency the ability to threaten a command and control style regulation to limit CO_2 unless the legislature passed a bill that superseded the EPA's authority.

21. Of course, for most nations (except perhaps the United States, in that its wealth and carbon consumption might make it what Olson would consider a large member), this also assumes that free riding problems are solved as well.

22. This, itself, represents an 8 percent increase since 2005, while worries about water and air pollution have actually declined.

23. There is a clear relationship between level of economic development and claimed knowledge about global warming. Of 127 nations surveyed, Japan leads the way at 99 percent and Tajikistan has the lowest claimed knowledge level at 15 percent (Pelham 2009a).

24. Specifically, respondents were asked, "How much do you know about global warming or climate change?" Those answering "something" or a "great deal" were classified as knowledgeable. Those so classified then responded to the following: "Temperature rise is a part of global warming or climate change. Do you think rising temperatures are . . . a result of human activities?"
25. Similarly, as there has been widespread trumpeting of the progress made in combating conventional air pollutants by members of both parties, as mentioned the percentage listing such ills as a major problem has declined noticeably.
26. These demands were as of June 2009 (for example Hille 2009) and reflected China's stance for quite some time.
27. Possibilities that may help counteract such resistance are requirements for substantial initial investments and the endogenous formation of domestic interests in favor of climate mitigation.
28. A nation's capacity is, admittedly, a broad and nebulous concept. A standard definition related to the environment comes from the Rio Summit, stating that it "encompasses a country's human, scientific, technological, organizational, institutional, and resource capabilities."
29. For example, deforestation, a major contributor to global warming, is a function of insecure property rights (for example Deacon 1999).
30. The clean development mechanism typically allows first world nations to reduce emissions in developing countries; because one must establish that the reductions are real and would not occur without the incentives offered, the mechanism is a ripe target for manipulation. Indeed, experience with its use as part of the Kyoto protocol has generated claims that most of the reductions—a major part of the European Union claiming that it has largely met its goals—are not real (Victor 2009).

REFERENCES

Aldy, Joseph E., Peter R. Orszag, and Joseph E. Stiglitz. 2001. Climate Change: An Agenda for Global Collective Action. Paper presented at the conference on The Timing of Climate Change Policies, Pew Center on Global Climate Change, Washington, October.

Aldy, Joseph E., Eduardo Ley, and Ian W.H. Parry. 2008. *A Tax-Based Approach to Slowing Global Climate Change*. Discussion Paper 08-26. Washington: Resources for the Future.

Anderson, Terry L., and Peter J. Hill. 1997. *Environmental Federalism*. Lanham, Maryland: Rowman and Littlefield.

Arrow, Kenneth J. 2007. Global Climate Change: A Challenge to Policy. *Economists Voice* 4 (3): article 2. Available at www.bepress.com/ev/vol4/iss3/art2.

Ball, Jeffrey. 2008. Currents: As Climate Heats Up, Questions of Cost Loom. *Wall Street Journal* (10 July): A10.

Baron, David P. 2001. Private Politics, Corporate Social Responsibility, and the Integrated Strategy. *Journal of Economics and Management Strategy* 10 (1): 7–45.

Barrett, Scott. 2007. *Why Cooperate? The Incentive to Supply Global Public Goods*. New York: Oxford University Press.

Barrett, Scott. 2009. Rethinking Global Climate Change Governance. *Economics: The Open-Access, Open-Assessment E-Journal* 3 (5). Available at www.economics-ejournal.org/economics/journalarticles/2009-5.

Baumert, Kevin A., Timothy Herzog, and Jonathan Pershing. 2005. *Navigating the Numbers: Greenhouse Gas Data and International Climate Policy*. Washington: World Resources Institute.

Bernauer, Thomas, and Vally Koubi. 2009. Effects of Political Institutions on Air Quality. *Ecological Economics* 68 (5): 1355–65.

Coase, Ronald. 1960. The Problem of Social Cost. *Journal of Law and Economics* 3 (1): 1–44.

Deacon, Robert T. 1999. Deforestation and Ownership: Evidence from Historical Accounts and Contemporary Data. *Land Economics* 75 (3): 341–59.

Dinda, Soumyanada. 2004. Environmental Kuznets Curve Hypothesis: A Survey. *Ecological Economics* 49 (4): 431–55.

Duval, Romain. 2008. *A Taxonomy of Instruments to Reduce Greenhouse Gas Emissions and Their Interactions*. OECD Economics Department Working Paper 636. Paris: Organisation for Economic Co-operation and Development.

Esty, Daniel C., and Maria H. Ivanova, editors. 2002. *Global Environmental Governance: Options and Opportunities*. New Haven, Connecticut: Yale School of Forestry and Environmental Studies.

Evans, Alex, and David Steven. 2009. *An Institutional Architecture for Climate Change: A Concept Paper*. New York: Center on International Cooperation, New York University.

Fischer, Carolyn, and Alan K. Fox. 2007. Output-Based Allocation of Emissions Permits for Mitigating Tax and Trade Interactions. *Land Economics* 83 (4): 575–99.

Garnaut, Ross. 2008. *The Garnaut Climate Change Review: Final Report*. New York: Cambridge University Press.

Glazer, Amihai, and Lawrence S. Rothenberg. 2001. *Why Government Succeeds and Why It Fails*. Cambridge, Massachusetts: Harvard University Press.

Goodell, Jeff. 2009. The Secretary of Saving the Planet. *Rolling Stone* 1081 (26 June): 59–65, 85–87.

Gore, Al. 2006. *An Inconvenient Truth: The Planetary Emergence of Global Warming and What We Can Do About It*. New York: Rodale.

Grossman, Gene M., and Alan B. Krueger. 1995. Economic Growth and the Environment. *Quarterly Journal of Economics* 110 (2): 353–77.

Hannesson, Rögnvaldur. 2004. *The Privatization of the Oceans*. Cambridge, Massachusetts: MIT Press.

Hansen, James, Makiko Sato, Pushker Kharecha, David Beerling, Robert Berner, Valerie Masson-Delmotte, Mark Pagani, Maureen Raymo, Dana L. Royer, and James C. Zachos. 2008. Target Atmospheric CO_2: Where Should Humanity Aim? *The Open Atmospheric Science Journal* 2:217–31.

Hepburn, Cameron. 2006. Regulation by Prices, Quantities, or Both: A Review of Instrument Choice. *Oxford Review of Economic Policy* 22 (2): 226–47.

Hille, Kathrin. 2009. Biggest Emitters Fail to Show the Rest of the World the Way Forward. *Financial Times* (11 June): 3.

Holtz-Eakin, Douglas, and Thomas M. Selden. 1995. Stoking the Fires? CO_2 Emissions and Economic Growth. *Journal of Public Economics* 57 (1): 85–101.

Hovi, Jon, Detlef F. Sprinz, and Arlid Underdal. 2009. Implementing Long-Term Climate Policy: Time Inconsistency, Domestic Politics, International Anarchy. *Global Environmental Policy* 9 (3): 20–39.

Inman, Robert P., and Daniel L. Rubinfeld. 1997. "The Political Economy of

Federalism." In *Perspectives on Public Choice: A Handbook*, edited by Dennis C. Mueller, pp. 73–105. Cambridge and New York: Cambridge University Press.

IPCC (Intergovernmental Panel on Climate Change). 2000. *IPCC Special Reports: Emissions Scenarios*. Geneva: Intergovernmental Panel on Climate Change. Available at www.ipcc.ch/pdf/special-reports/spm/sres-en.pdf.

IPCC. 2007. *Climate Change 2007: Synthesis Report*. Geneva: Intergovernmental Panel on Climate Change. Available at www.ipcc.ch/pdf/assessment-report/ar4/syr/ar4_syr.pdf.

Jacobsen, Mark R. 2008. Evaluating U.S. Fuel Economy Standards in a Model with Producer and Household Heterogeneity. Unpublished paper; Department of Economics, University of California, San Diego.

Keohane, Robert O., and Karl Raustiala. 2008. Toward a Post-Kyoto Climate Change Architecture: A Political Analysis. Unpublished paper; Woodrow Wilson School of Public and International Affairs, Princeton University.

Keohane, Nathaniel O., Richard L. Revesz, and Robert N. Stavins. 1999. "The Positive Political Economy of Instrument Choice in Environmental Policy." In *Environmental and Public Economics: Essays in Honor of Wallace Oates*, edited by Arvind Panagariya, Paul Portney, and Robert Schwab. Cheltenham, UK and Northampton, MA, USA: Edward Elgar, pp. 89–125.

Kirchgassner, Gebhard, and Friedrich Schneider. 2003. On the Political Economy of Environmental Policy. *Public Choice* 115 (3): 369–96.

Kolk, Ans, and Jonatan Pinkse. 2007. Multinationals' Political Activities on Climate Change. *Business and Society* 46 (2): 201–28.

Lai, Yu-Bong. 2008. Auctions or Grandfathering: The Political Economy of Tradable Emissions Permits. *Public Choice* 136 (1–2): 181–200.

Lewis-Beck, Michael S., and Martin Paldam. 2000. Economic Voting: An Introduction. *Electoral Studies* 19 (2–3): 113–21.

Lohmann, Susanne. 2003. Representative Government and Special Interest Politics (We Have Met the Enemy and He Is Us). *Journal of Theoretical Politics* 15 (3): 299–319.

Mandell, Svante. 2008. Optimal Mix of Emissions Taxes and Cap-and-Trade. *Journal of Environmental Economics and Management* 56 (2): 131–40.

McKibben, Bill. 2007. Remember This: 350 Parts per Million. *Washington Post* (28 December): A21.

Michaelowa, Axel, and Frank Jotzo. 2005. Transaction Costs, Institutional Rigidities, and the Clean Development Mechanism. *Energy Policy* 33 (4): 511–23.

Murray, Brian R., Richard G. Newell, and William A. Pizer. 2008. *Balancing Costs and Emissions Certainty: An Allowance Reserve for Cap-and-Trade*. Discussion paper RFF DP 08-24. Washington: Resources for the Future.

Newell, Richard G., and William A. Pizer. 2003. Discounting the Distant Future: How Much do Uncertain Rates Increase Valuations? *Journal of Environmental Economics and Management* 46 (1): 52–71.

Newell, Richard G., James N. Sanchirico, and Suzi Kerr. 2005. Fishing Quota Markets. *Journal of Environmental Economics and Management* 49 (2): 437–62.

Nisbet, Matthew C., and Teresa Myers. 2007. The Polls—Trends: Twenty Years of Public Opinion about Global Warming. *Public Opinion Quarterly* 71 (3): 444–70.

Nordhaus, William D. 2001. Global Warming Economics. *Science* 294 (5545): 1283–84.

Nordhaus, William D. 2007. To Tax or Not to Tax: Alternative Approaches to Slowing Global Warming. *Review of Environmental Economics and Policy* 1 (1): 26–44.

Nordhaus, William D. 2009. Economic Issues in Designing a Global Agreement on Global Warming. Keynote address at the conference on Climate Change: Global Risks, Challenges, and Decisions, Copenhagen, 10–12 March 2009.

Oates, Wallace E., and Paul R. Portney. 2003. "The Political Economy of Environmental Policy." In *Handbook of Environmental Economics*, edited by Karl-Goran Maler and Jeffrey R. Vincent, volume 1. Amsterdam: Elsevier Science.

Oates, Wallace E., and Robert M. Schwab. 1988. Economic Competition among Jurisdictions: Efficiency Enhancing or Distortion Inducing. *Journal of Public Economics* 35 (3): 333–54.

Olmstead, Sheila M., and Robert N. Stavins. 2006. An International Policy Architecture for the Post-Kyoto Era. *American Economic Review Papers and Proceedings* 96 (2): 35–38.

Olson, Jr., Mancur. 1965. *The Logic of Collective Action: Public Goods and the Theory of Groups.* Cambridge, Massachusetts: Harvard University Press.

Olson, Jr., Mancur, and Richard Zeckhauser. 1966. An Economic Theory of Alliances. *Review of Economics and Statistics* 48 (3): 266–79.

Oreskes, Naomi. 2004. The Scientific Consensus on Climate Change. *Science* 308 (5702): 1686.

Pacala, S., and R. Socolow. 2004. Stabilization Wedges: Solving the Climate Problem for the Next 50 Years with Current Technologies. *Science* 305:968–72.

Pelham, Brett. 2009a. Awareness, Opinion about Global Warming Vary Worldwide: Many Unaware, Do not Necessarily Blame Human Activities. *Gallup Poll* (22 April).

Pelham, Brett. 2009b. Views on Global Warming Related to Energy Efficiency: Relationship Exists Regardless of National Wealth, Literacy. *Gallup Poll* (24 April).

Pew Research Center. 2008. *A Deeper Partisan Divide over Global Warming.* Survey Report (8 May). Washington: Pew Research Center. Available at people-press.org/reports.

Putnam, Robert D. 1988. Diplomacy and Domestic Politics: The Logic of Two-Level Games. *International Organization* 42 (3): 427–60.

Rothenberg, Lawrence S. 2002. *Environmental Choices: Policy Responses to Green Demands.* Washington: Congressional Quarterly Press.

Saad, Lydia. 2009. Water Pollution Americans' Top Green Concern: Worry about Environmental Problems Has Edged up since 2004. *Gallup Poll* (25 March).

Samuelson, Paul A. 1954. The Pure Theory of Public Expenditure. *Review of Economics and Statistics* 36 (4): 387–89.

Sandler, Todd, and Keith Hartley. 2001. Economics of Alliances: The Lessons for Collective Action. *Journal of Economic Literature* 39 (3): 869–96.

Sandvik, Hanno. 2008. Public Concern over Global Warming Correlates Negatives with National Wealth. *Climate Change* 90 (3): 333–41.

Schmalensee, Richard, Thomas M. Stoker, and Ruth A. Judson. 2004. World Carbon Dioxide Emissions: 1950–2050. *The Review of Economics and Statistics* 80 (1): 15–27.

Socolow, Robert H., and Stephen W. Pacala. 2006. A Plan to Keep Carbon in Check. *Scientific American* 295 (3): 50–57.

Spruiell, Stephen, and Kevin Williamson. 2009. A Garden of Piggish Delights. *National Review Online* (2 July).

Stavins, Robert N. 2004a. Forging a More Effective Climate Change Treaty: Can the Next Treaty be Based on Sound Science, Rational Economics, and Pragmatic Politics? *Environment* 46 (10): 23–30.

Stavins, Robert N. 2004b. *The Political Economy of Environmental Regulation*. Cheltenham, UK and Northampton, MA, USA: Edward Elgar.

Stern, David I. 2004. The Rise and Fall of the Environmental Kuznets Curve. *World Development* 32 (8): 1419–39.

Stern, Nicholas. 2008. The Economics of Climate Change. *American Economic Review: Papers and Proceedings* 98 (2): 1–37.

Stiglitz, Joseph E. 2006. A New Agenda for Global Warming. *Economist's Voice* 3 (7): article 3. Available at www.bepress.com/ev/vol3/iss7/art3.

Sunstein, Cass R. 2005. *Laws of Fear: Beyond the Precautionary Principle*. New York: Cambridge University Press.

Sunstein, Cass R. 2007. On Discounting Regulatory Benefits: Risk, Money, and Intergenerational Equity. *University of Chicago Law Review* 74 (1): 171–208.

Sunstein, Cass R., and Arden Rowell. 2007. Of Montreal and Kyoto: A Tale of Two Protocols. *Harvard Environmental Law Review* 31 (1): 1–65.

Tol, Richard S.J. 2008. The Social Costs of Carbon: Trends, Outliers, and Catastrophes. *Economics: The Open-Access, Open-Assessment E-Journal* 2 (25): 1–24.

United Nations. 2009. *The State of World Fisheries and Aquaculture 2008*. Rome: Food and Agriculture Organization of the United Nations.

United Nations Framework Convention on Climate Change (UNFCC). 2007. *Climate Change: Impacts, Vulnerabilities and Adaptation in Developing Countries*. Bonn: Climate Change Secretariat.

Victor, David G. 2009. *Global Warming Policy after Kyoto: Rethinking Engagement with Developing Countries*. Working Paper 82. Stanford, California: Stanford Program on Energy and Sustainable Development, Stanford University.

Wall Street Journal. 2009. The Cap and Tax Fiction: Democrats Off-loading Economics to Pass Climate Change Bill. *Wall Street Journal* (June 26).

Weitzman, Martin L. 1974. Prices versus Quantities. *Review of Economic Studies* 41 (4): 477–91.

Zaller, John. 1992. *The Nature and Origins of Mass Opinion*. New York: Cambridge University Press.

6. Climate change meets trade in promoting green growth: potential conflicts and synergies

ZhongXiang Zhang

INTRODUCTION

Climate and trade policies both affect the use of natural resources. Their linkages are recognized in the objectives of the corresponding agreements to safeguard the two regimes. The ultimate objective of the United Nations Framework Convention on Climate Change (UNFCCC) is to stabilize greenhouse gas concentrations in the atmosphere. An underlying principle to guide this effort is that "measures taken to combat climate change, including unilateral ones, should not constitute a means of arbitrary or unjustifiable discrimination or a disguised restriction on international trade." At the same time, the World Trade Organization (WTO) Agreement recognizes that trade should be conducted "while allowing for the optimal use of the world's resources in accordance with the objective of sustainable development, seeking both to provide and preserve the environment and to enhance the means for doing so."

Clearly, the main aim of both the UNFCCC and the WTO is to ensure efficiency in the use of resources, from the perspective of either maximizing the gains from the comparative advantage of nations through trade or ensuring that economic development is environmentally sustainable. Therefore, the objectives of the UNFCCC (and its Kyoto protocol) and the WTO are not explicitly in conflict with each other.

However, the possibility of conflicts may arise in implementing the Kyoto protocol and any international regime to succeed it as countries aim for green growth. With greenhouse gas emissions embodied in virtually all products produced and traded in every conceivable economic sector, effectively addressing climate change will require a fundamental transformation of our economy and the ways energy is produced and used. This will certainly have a bearing on world trade, because it will affect the costs of production of traded products and therefore their competitive positions in the world market. This climate-trade nexus has become the focus of an academic debate (for example Bhagwati and Mavroidis 2007; Charnovitz 2003; Ismer and Neuhoff 2007; Swedish National Board of Trade 2004; World Bank 2007; Zhang 1998, 2004 and 2007a; Zhang and Assunção 2004), and is gaining increasing attention as governments are taking great efforts to implement the Kyoto protocol and forge a post-2012 climate change regime to succeed it.

To comply with the Kyoto protocol, Annex I countries are preparing, adopting, and implementing comprehensive measures to meet their emissions targets set under the protocol. The Kyoto protocol gives these countries considerable flexibility in the choice of domestic policies to meet their emissions commitments. Possible climate measures include carbon and energy taxes, subsidies, energy efficiency standards, eco-labels, government procurement policies, and flexibility mechanisms built into the Kyoto protocol. The implementation of these measures has the potential to affect trade and thus raises concerns about compatibility with WTO rules.

In order to meet their Kyoto emissions targets or stimulate their economies with minimum adverse effects, or both, it is very likely that Annex I governments with differentiated legal and political systems might pursue emission reduction policies in such a way as to favor domestic producers over foreign ones. Such differential treatment could occur in governing eligibility for, and the amount of, a subsidy, in establishing energy efficiency standards, in determining the category of eco-labeled products and the procedures for establishing eco-labels, and in specifying criteria for tenders and condition for participating in government procurement bids as "Buy American" type of provisions biases for U.S. home-made goods under its stimulus package. In the case where a country unilaterally imposes a carbon tax or a cap-and-trade regime, it may adjust taxes or carbon contents of traded products at the border to mitigate competitiveness effects of cheaper imports not subject to a similar level of the carbon tax or emission limits in the country of origin. Measures of this sort raise complex questions with respect to their WTO consistency and the conditions under which border taxes or border carbon contents of traded products can be adjusted to accommodate a loss of international competitiveness.[1]

This climate-trade nexus becomes intensive as countries are developing post-2012 climate commitments on the basis of the "Bali Road Map," which was agreed to at the UNFCCC Conference of Parties meeting in December 2007, with a clear deadline for conclusion by 2009 at Copenhagen. No one would disagree that the U.S. commitment to cut emissions is essential to such a global pact and President Obama's desire to lead after what is viewed as eight years of lost time under President Bush.

However, much of Obama's ability to move forward in international climate negotiations rests with the U.S. Congress, because the Obama administration will likely be in the position to agree to a specific emissions target that the whole world has long awaited only when Congress has enacted or is on the verge of enacting legislation capping U.S. greenhouse gas emissions.

The Intergovernmental Panel on Climate Change calls for developed countries to cut their greenhouse gas emissions by 25–40 percent by 2020 and by 80 percent by 2050 relative to their 1990 levels, in order to avoid dangerous climate change impacts. In the meantime, under the UNFCCC principle of "common but differentiated responsibilities," developing countries are allowed to move at different speeds, as do their developed counterparts. This principle is clearly reflected in the Bali roadmap, which requires developing countries to take "nationally appropriate mitigation actions . . . in the context of sustainable development, supported and enabled by technology, financing and capacity-building, in a measurable, reportable and verifiable manner." Understandably, the United States and other industrialized countries would like to see developing countries, in particular large developing economies, go beyond that because of concerns about their own competitiveness and growing greenhouse gas emissions in developing countries. They are considering unilateral trade measures to "induce" developing countries to do so. This has been the case in the course of debating and voting the U.S. congressional climate bills capping U.S. greenhouse gas emissions. U.S. legislators have pushed for major emerging economies, such as China and India, to take climate actions comparable to those of the United States. Otherwise, their products sold on the U.S. market will have to purchase and surrender emissions allowances to cover their carbon contents. Border carbon adjustment measures of this kind have raised great concern about whether they are WTO-consistent and have received heavy criticism from developing countries.

To date, border adjustment measures in the form of emissions allowance requirements (EAR) under the U.S. proposed cap-and-trade regime are the most concrete unilateral trade measure put forward to level the carbon playing field. If improperly implemented, such measures could disturb the world trade order and trigger a trade war. Because of these potentially far-reaching impacts, this chapter focuses on this type of unilateral border adjustment. It requires importers to acquire and surrender emissions allowances corresponding to the embedded carbon contents in their goods from countries that have not taken climate actions comparable to those of the home country. The discussion here is mainly on the legality of unilateral EAR under the WTO rules.[2]

The next section briefly describes the border carbon adjustment measures proposed in the U.S. legislation. The third section deals with the WTO scrutiny of EAR proposed in the U.S. congressional climate bills. The fourth section briefly discusses whether an EAR threat would be effective as an inducement for major emerging economies to take climate actions that they would otherwise not take, as well as methodological challenges in implementing EAR. The chapter ends with some concluding remarks on the needs on the U.S. side to minimize the potential conflicts with WTO provisions in designing such border carbon adjustment measures, and with suggestions for major developing countries being targeted by such border measures to deal effectively with the proposed border adjustment measures to their advantage.

PROPOSED BORDER ADJUSTMENT MEASURES IN U.S. LEGISLATION

The notion of border carbon adjustments (BCA) is not an American invention. The idea of using BCA to address the competitiveness concerns as a result of differing climate policies was first floated in the European Union, in response to the U.S. withdrawal from the Kyoto protocol. Dominique de Villepin, the French prime minister, proposed in November 2006 for carbon tariffs on goods from countries that had not ratified the Kyoto protocol. He clearly had the United States in mind when contemplating such proposals aimed to bring the United States to the table for climate negotiations. However, Peter Mandelson, the E.U. trade commissioner, dismissed the French proposal as not only a probable breach of trade rules but also "not good politics" (Bounds 2006). As a balanced reflection of the divergent views on this issue, the European Commission has suggested that it could implement a "carbon equalization system . . . with a view to putting E.U. and non-E.U. producers on a comparable footing. . . . Such a system could apply to importers of goods requirements similar to those applicable to installations within the European Union, by requiring the surrender of allowances" (European Commission 2008). While the European Union has considered the possibility of imposing a border allowance adjustment, should serious leakage issues arise in the future, it has put this option on hold at least until 2012. The European Commission has proposed using temporary free allocations to address competitiveness concerns in the interim. Its aim is to facilitate a post-2012 climate negotiation while keeping that option in the background as a last resort.

Interestingly, the U.S. legislators have not only embraced such BCA measures (which they had been opposed to) but have also focused on their design issues in more detail. In the U.S. Senate, the Boxer Substitute of the Lieberman-Warner Climate Security Act (S. 3036) mandates that starting from 2014 importers of products covered by the cap-and-trade scheme would have to purchase emissions allowances from an International Reserve Allowance Program if no comparable climate action were taken in the exporting country. Least developed countries and countries that emit less than 0.5 percent of global greenhouse gas emissions (that is, those considered not significant emitters) would be excluded from the scheme. Given that most carbon-intensive industries in the United States run a substantial trade deficit (Houser et al. 2008), this proposed EAR clearly aims to level the carbon playing field for domestic producers and importers. In the U.S. House of Representatives, the American Clean Energy and Security Act of 2009 (H.R. 2998),[3] sponsored by Representatives Henry Waxman (D-CA) and Edward Markey (D-MA), was narrowly passed in June 2009. The so-called Waxman-Markey bill set up an International Reserve Allowance Program in which U.S. importers of primary emission-intensive products would be required to acquire and surrender carbon emissions allowances, if the imports come from countries that have not taken "greenhouse gas compliance obligations commensurate with those that would apply in the United States." The European Union by any definition would pass this comparability test, because it has undertaken the Kyoto protocol and is going to take in its follow-up regime much more ambitious climate targets than the United States. Because all remaining Annex I countries except the United States have accepted mandatory emissions targets under the Kyoto protocol, these countries would likely pass the comparability test as well, which exempts them from EAR under the U.S. proposed cap-and-trade regime. While France targeted American goods, the U.S. EAR clearly targets major emerging economies, such as China and India.

WTO SCRUTINY OF U.S. CONGRESSIONAL CLIMATE BILLS

The import emissions allowance requirement was a key part of the Lieberman-Warner Climate Security Act of 2008, and will reappear as the U.S. Senate starts writing, debating, and voting its own version of a climate change bill after the U.S. House of Representatives narrowly passed the Waxman-Markey bill. Moreover, concerns raised in the Lieberman-Warner bill seem to have provided references to writing relevant provisions in the Waxman-Markey bill to deal with the competitiveness concerns. For these reasons, the following discussion begins with the Lieberman-Warner bill.

A proposal first introduced by the International Brotherhood of Electrical Workers and American Electric Power in early 2007 would require importers to acquire emissions allowances to cover the carbon content of certain products from countries that do not take climate actions comparable to those of the United States (Morris and Hill 2007). The original version of the Lieberman-Warner bill incorporated this mechanism, threatening to punish energy-intensive imports from developing countries by requiring importers to obtain emissions allowances, but only if they had not taken comparable actions by 2020, eight years after the effective start date of a U.S. cap-and-trade regime begins. It was argued that the inclusion of trade provisions would give the United States additional diplomatic leverage to negotiate multilaterally and bilaterally with other countries on comparable climate actions. Should such negotiations not succeed, trade provisions would provide a means of leveling the carbon playing field between American energy-intensive manufacturers and their competitors in countries not taking comparable climate actions. Not only would the bill have imposed an import allowance purchase requirement too quickly, it would have also dramatically expanded the scope of punishment: almost any manufactured product would potentially have qualified. If strictly implemented, such a provision would pose an insurmountable hurdle for developing countries (*Economist* 2008).

It should be emphasized that the aim of including trade provisions is to facilitate negotiations while keeping open the possibility of invoking trade measures as a last resort. The latest version of the Lieberman-Warner bill has brought the deadline forward to 2014 to gain business and union backing.[4] The inclusion of trade provisions might be considered the "price" of passage for any U.S. legislation capping its greenhouse gas emissions. Put another way, it is likely that no climate legislation can move through the U.S. Congress without dealing with the issue of trade provisions. An important issue on the table is the length of the grace period to be granted to developing countries. While many factors need to be taken into consideration here (Haverkamp 2008), further bringing forward the imposition of allowance requirements to imports is rather unrealistic, given the already very short grace period ending in 2019 in the original version of the bill. It should be noted that the Montreal protocol on Substances that Deplete the Ozone Layer grants developing countries a grace period of ten years (Zhang 2000). Given that the scope of economic activities affected by a climate regime is several orders of magnitude larger than those covered by the Montreal protocol, if legislation incorporates border adjustment measures (setting the issue of their WTO consistency aside), in my view, they should not be invoked for at least ten years after mandatory U.S. emissions targets take effect.

Moreover, unrealistically shortening the grace period granted before resorting to the trade provisions would increase the uncertainty of whether the measure would withstand a challenge by U.S. trading partners before the WTO. As the ruling in the shrimp-turtle dispute indicates (see Box 6.1), for a trade measure to be considered WTO-consistent, a period of good-faith efforts to reach agreements among the countries concerned is needed before imposing such trade measures. Put another way, trade provisions should be preceded by major efforts to negotiate with partners within a reasonable time frame. Furthermore, developing countries need reasonable time to develop and operate national climate policies and measures. Take the establishment of an emissions trading scheme as a case in point. Even for the U.S. SO_2 Allowance Trading Program, the entire process from the time when the U.S. Environmental Protection Agency began to compile the data for its allocation database in 1989 to the publication of its final allowance allocations in March 1993 took almost four years. For the first phase of the E.U. Emissions Trading Scheme, the entire process took almost two years, from the publication of the E.U. directive establishing a scheme for greenhouse gas emission allowance trading in July 2003 to the approval of the last national allocation plan (for Greece) in June 2005. For developing countries that have very weak environmental institutions and do not have dependable data on emissions, fuel uses, and outputs for installations, this allocation process is expected to take much longer than what was experienced in the United States and the European Union (Zhang, 2007b).

In the case of a WTO dispute, the question will arise whether there are any alternatives to trade provisions that could be reasonably expected to fulfill the same function but are not inconsistent or are less inconsistent with the relevant WTO provisions. Take the General Agreement on Tariffs and Trade (GATT) Thai cigarette dispute as a case in point. Under Section 27 of the Tobacco Act of 1966, Thailand restricted imports of cigarettes and imposed a higher tax rate on imported cigarettes when they were allowed on the three occasions since 1966: namely in 1968–70, 1976, and 1980. After consultations with Thailand failed to lead to a solution, the United States asked the Dispute Settlement Panel in 1990 to rule on the Thai action, on the grounds that it was inconsistent with Article XI:1 of the General Agreement; was not justified by the exception under Article XI:2(c), because cigarettes were not an agricultural or fisheries product in the meaning of Article XI:1; and was not justified under Article XX(b) because the restrictions were not necessary to protect human health: in other words, controlling the consumption of cigarettes did not require an import ban. The Dispute Settlement Panel ruled against Thailand. The Panel found that Thailand had acted inconsistently with Article

BOX 6.1   IMPLICATIONS OF THE WTO FINDINGS
                ON THE SHRIMP-TURTLE DISPUTE

To address the decline of sea turtles around the world, in 1989
the U.S. Congress enacted Section 609 of Public Law 101-162
to authorize embargoes on shrimp harvested with commercial
fishing technology harmful to sea turtles. The United States
was challenged in the WTO by India, Malaysia, Pakistan, and
Thailand in October 1996, after embargoes were leveled against
them. The four governments challenged this measure, asserting
that the United States could not apply its laws to foreign process
and production methods. A WTO Dispute Settlement Panel was
established in April 1997 to hear the case. The panel found that
the United States failed to approach the complainant nations
in serious multilateral negotiations before enforcing the U.S.
law against those nations. The panel held that the U.S. shrimp
embargo was a class of measures of processes-and-production
methods type and had a serious threat to the multilateral trading
system because it conditioned market access on the conservation policies of foreign countries. Thus, it could not be justified
under GATT Article XX.

   However, the WTO Appellate Body overruled the panel's
reasoning. The Appellate Body held that a WTO member requiring from exporting countries compliance, or adoption of, certain
policies prescribed by the importing country does not render the
measure inconsistent with the WTO obligation. Although the
Appellate Body still found that the U.S. shrimp embargo was justified under GATT Article XX, the decision was not on a ground
that the U.S. sea turtle law itself was inconsistent with GATT.
Rather, the ruling was on a ground that the application of the law
constituted "arbitrary and unjustifiable discrimination" between
WTO members (WTO 1998). The WTO Appellate Body pointed
to a 1996 regional agreement reached at U.S. initiation, namely
the Inter-American Convention on Protection and Conservation
of Sea Turtles, as evidence of the feasibility of such an approach
(WTO 1998; Berger 1999). Here, the Appellate Body again
advanced the standing of multilateral environmental treaties
(Zhang 2004; Zhang and Assunção 2004). Thus, the ruling on
this trade dispute under the WTO could be interpreted as a clear
preference for actions taken pursuant to multilateral agreements

and/or negotiated through international cooperative arrangements, such as the Kyoto Protocol and its successor. However, this interpretation should be drawn with great caution, because there is no doctrine of stare decisis (namely, "to stand by things decided") in the WTO; the GATT/WTO panels are not bound by previous panel decisions (Zhang and Assunção 2004).

Moreover, the WTO shrimp-turtle dispute settlement has a bearing on the ongoing discussion on the "comparability" of climate actions in a post-2012 climate change regime. The Appellate Body found that when the United States shifted its standard from requiring measures essentially the same as the U.S. measures to "the adoption of a program comparable in effectiveness," this new standard would comply with the WTO disciplines (WTO 2001, paragraph 144). Some may view this case as opening the door for U.S. climate legislation that bases trade measures on an evaluation of the comparability of climate actions taken by other trading countries. Comparable action can be interpreted as meaning action comparable in effect as the "comparable in effectiveness" in the shrimp-turtle dispute. It can also be interpreted as meaning "the comparability of efforts." The Bali Action Plan adopts the latter interpretation, using the term comparable as a means of ensuring that developed countries undertake commitments comparable to each other (Zhang 2009a).

XI:1 for having not granted import licenses over a long period of time. Recognizing that XI:2(c) allows exceptions for fisheries and agricultural products—if the restrictions are necessary to enable governments to protect farmers and fishermen (who, because of the perishability of their produce, often could not withhold excess supplies of the fresh product from the market)—the panel found that cigarettes were not "like" the fresh product as leaf tobacco and thus were not among the products eligible for import restrictions under Article XI:2(c). Moreover, the panel acknowledged that Article XX(b) allowed contracting parties to give priority to human health over trade liberalization. The panel held the view that the import restrictions imposed by Thailand could be considered "necessary" in terms of Article XX(b) only if there were no alternative measure consistent with the General Agreement, or less inconsistent with it, which Thailand could reasonably be expected to employ to achieve its health policy objectives. However, the panel found the Thai import restriction measure not necessary, because Thailand could reasonably be expected to adopt strict, nondiscriminatory labelling and ingredient disclosure regulations and to ban all the direct and indirect advertising, promotion, and sponsorship of cigarettes to ensure the quality and reduce the quantity of cigarettes sold in Thailand. These alternative measures are considered WTO-consistent to achieve the same health policy objectives that Thailand was pursuing through an import ban on all cigarettes whatever their ingredients (GATT 1990). Simply put, in the GATT Thai cigarette dispute, the Dispute Settlement Panel concluded that Thailand had legitimate concerns with health but had measures available to it other than a trade ban that would be consistent with the General Agreement on Tariffs and Trade: for example, bans on advertising (GATT 1990).

Indeed, there are alternatives to resorting to trade provisions to protect the U.S. trade-sensitive, energy-intensive industries during a period when the United States is making good-faith efforts to negotiate with trading partners on comparable actions. One way to address competitiveness concerns is to initially allocate free emission allowances to those sectors vulnerable to global competition, either totally or partially.[5] Bovenberg and Goulder (2002) found that giving out about 13 percent of the allowances to fossil fuel suppliers freely instead of auctioning in an emissions trading scheme in the United States would be sufficient to prevent their profits with the emissions constraints from falling in comparison with those without the emissions constraints.

There is no disagreement that the allocation of permits to emissions sources is a politically contentious issue. Grandfathering, or at least partially grandfathering, helps these well-organized, politically highly-mobilized industries or sectors to save considerable expenditure and thus increases the political acceptability of an emissions trading scheme, although it leads to a higher economic cost than a policy where the allowances are fully auctioned.[6] This explains why the sponsors of the American Clean Energy and Security Act of 2009 had to make a compromise amending the act to auction only 15 percent of the emission permits instead of the initial proposal for auctioning all of them in a proposed cap-and-trade regime. This change allowed the U.S. House of Representatives Energy and Commerce Committee to secure passage of the act in May 2009. However, it should be pointed out that although grandfathering is thought of as giving implicit subsidies to these sectors, grandfathering is less trade-distorted than exemptions from carbon taxes (Zhang 1998 and 1999), which means that partially grandfathering is even less trade-distorted than exemptions from carbon taxes. To understand the difference, it is important to bear in mind that grandfathering itself also implies an opportunity cost for firms receiving permits: what matters here is not how firms get permits, but what firms can sell them for. That is what determines opportunity cost. Thus, even if permits are awarded gratis, firms will value them at their market price. Accordingly, the prices of energy will adjust to reflect the increased scarcity of fossil fuels. This means that regardless of whether emissions permits are given out freely or are auctioned by the government, the effects on energy prices are expected to be the same, although the initial ownership of emissions permits differs among different allocation methods. As a result, relative prices of products will not be distorted relative to their pre-existing levels, and switching of demand toward products of those firms whose permits are awarded gratis (the so-called substitution effect) will not be induced by grandfathering. This makes grandfathering different from the exemptions from carbon taxes. In the latter case, substitution effects exist (Zhang 1998 and 1999). For example, the Commission of the European Communities (CEC) proposal for a mixed carbon and energy tax[7] provides for exemptions for the six energy-intensive industries (iron and steel, non-ferrous metals, chemicals, cement, glass, and pulp and paper) from coverage by the CEC tax on grounds of competitiveness. This not only reduces the effectiveness of the CEC tax in achieving its objective of reducing CO_2 emissions, but also makes industries that are exempt from paying the CEC tax improve their competitive position in relation to industries that are not. Therefore, there will be some switching of demand toward the products of these energy-intensive industries, which is precisely the reaction that such a tax should avoid (Zhang 1997).

The import allowance requirement approach would distinguish between two otherwise physically identical products on the basis of climate actions in place in the country of origin. This discrimination of like products among trading nations would constitute a prima facie violation of WTO rules. To pass WTO scrutiny of trade provisions, the United States is likely to make reference to the health and environmental exceptions provided under GATT Article XX (see Box 6.2). This article itself is the exception that authorizes governments to employ otherwise GATT-illegal measures when such measures are necessary to deal with certain enumerated public policy problems. The GATT panel in Tuna/Dolphin II concluded that Article XX does not preclude governments from pursuing environmental concerns outside their national territory, but such extrajurisdictional application of domestic laws would be permitted only if aimed *primarily* (emphasis added) at having a conservation or protection effect (GATT 1994; Zhang 1998). The capacity of the planet's atmosphere to absorb greenhouse gas emissions without adverse impacts is an "exhaustible natural resource." Thus, if countries take measures on their own, including extrajurisdictional application primarily to prevent the depletion of this "exhaustible natural resource," such measures will have a good

---

BOX 6.2   CORE WTO PRINCIPLES

GATT Article 1 ("most favored nation" treatment): WTO members not allowed to discriminate against like imported products from other WTO members.

GATT Article III ("national treatment"): Domestic and like imported products treated identically, including any internal taxes and regulations.

GATT Article XI ("elimination of quantitative restrictions"): Forbids any restrictions (on other WTO members) in the form of bans, quotas, or licenses.

GATT Article XX:

> Subject to the requirement that such measures are not applied in a manner which would constitute a means of arbitrary or unjustifiable discrimination between countries where the same conditions prevail, or a disguised restriction on international trade, nothing in this Agreement shall be constructed to prevent the adoption or enforcement by any contracting party of measures . . .
> (b) necessary to protect human, animal or plant life or health . . .
> (g) relating to the conservation of exhaustible natural resources if such measures are made effective in conjunction with restrictions on domestic production or consumption. . . .

The threshold for (b) is higher than for (g), because, in order to fall under (b), the measure must be "necessary," rather than merely "relating to" under (g).

---

justification under GATT Article XX. Along this reasoning, if the main objective of trade provisions is to protect the environment by requiring other countries to take actions comparable to those of the United States, then mandating importers to purchase allowances from the designated special international reserve allowance pool to cover the carbon emissions associated with the manufacture of a product is debatable. To increase the prospects for a successful WTO defense, I think that trade provisions can refer to the designated special international reserve allowance pool, but may not do so without adding "or equivalent." This will allow importers to submit equivalent emissions reduction units that are not necessarily allowances but are recognized by international treaties to cover the carbon contents of imported products.

Clearly, these concerns raised in the Lieberman-Warner bill have shaped relevant provisions in the Waxman-Markey bill to deal with the competitiveness and leakage concerns. Accordingly, the Waxman-Markey bill has avoided all the aforementioned controversies raised in the Lieberman-Warner bill. Unlike the EAR in the Lieberman-Warner bill—which focuses exclusively on imports into the United States, but does nothing to address the competitiveness of U.S. exports in foreign markets—the Waxman-Markey bill included both rebates for a few energy-intensive, trade-sensitive sectors[8] and free emissions allowances to avoid putting U.S. manufacturers at a disadvantage relative to overseas competitors. Unlike the Lieberman-Warner bill in the U.S. Senate, the Waxman-Markey bill also gives China, India, and other major developing nations time to enact their climate-friendly measures. Under the Waxman-Markey bill, the International Reserve Allowance Program may not begin before January 2025. The U.S. President may implement an International Reserve Allowance Program only for sectors producing primary products. While the bill called for a "carbon tariff" on imports, it very much framed that measure as a last resort that a U.S. President could impose at his or her discretion regarding border adjustments or tariffs. However, in the middle of the night before the vote on 26 June 2009, a provision was inserted in this House bill that requires the President, starting in 2020, to impose a border adjustment—or tariffs—on certain goods from countries that do not act to limit their greenhouse gas emissions. The President can waive the tariffs only if explicit permission is received from the U.S. Congress (Broder 2009). The last-minute changes in the bill changed a presidential long-term back-up option to a requirement that the President put such tariffs in place under specified conditions. Such changes significantly altered the spirit of the bill, moving it considerably closer to risky protectionism. While praising the passage of the House bill as an "extraordinary first step," President Obama opposed a trade provision in that bill.[9] The carbon tariff proposals have also drawn fierce criticism from China and India. Without specific reference to the United States or the Waxman-Markey bill, China's Ministry of Commerce said in a statement posted on its website that proposals to impose "carbon tariffs" on imported products will violate the rules of the World Trade Organization. That would enable developed countries to "resort to trade in the name of protecting the environment." The carbon tariff proposal runs against the principle of "common but differentiated responsibilities," the spirit of the Kyoto protocol. This will neither help to strengthen confidence that the international community can cooperate to handle the economic crisis, nor help any country's endeavors during climate change negotiations. Thus China is strongly opposed to it (Ministry of Commerce 2009).

INEFFECTIVE INDUCEMENT AND METHODOLOGICAL CHALLENGES

Proponents of an EAR argue that such a threat would be effective as an inducement for major emerging economies to take on such a level of climate action at which U.S. legislation aims. However, this is questionable. The EAR under the U.S. proposed cap-and-trade regime would not apply to all imports. Rather, it would specifically target primary emissions-intensive products, such as steel, aluminum, and cement. Indeed, China has become a key producer of these primary products, accounting for 36 percent of global steel production, 32 percent of global aluminum production and over 50 percent of global cement production in 2007. The logic for the threat of EAR is that the fear of losing market access for these products would be enough to jawbone China to take climate actions that it otherwise would not take. However, the problem with this logic is that China's burgeoning supply of these carbon-intensive products is not mainly destined for export. Rather, they are made in China for China, going primarily to meet China's own demand. As the world's largest steel exporter, China exported only 2 percent of its steel production to the European Union and less than 1 percent to the United States in 2007. As the world's largest cement producer and exporter, China consumed 97 percent of its cement domestically and exported less than 1 percent of its production to the United States in 2007 (Houser 2008; Houser et al. 2008). Even if an EAR were implemented jointly with the European Union, it would have little leverage effect on China, because China is unlikely to raise the cost of producing 97 percent of its output (destined for the domestic market) in order to protect a market of less than 3 percent of its production (destined abroad). Moreover, this effect on the targeted country will be further alleviated by rerouting trade flows to deliver the covered products from countries that are not subject to the EAR scheme. With Japan passing the comparability test and thus being exempted from an EAR under the proposed U.S. cap-and-trade regime, imposing an EAR on Chinese steel, but not on Japanese steel, could make Japanese steel more competitive in the U.S. market than Chinese steel. That could lead Japanese steel makers to sell more steel to the United States and Japanese steel consumers to import more from China (Houser et al. 2008). In the end, this neither affects China nor protects U.S. steel producers.

Besides the issue of WTO consistency and the ineffectiveness of an EAR in leveraging developing countries to change behavior, there will be methodological challenges in implementing an EAR under a cap-and-trade regime, although such practical implementation issues are secondary concerns. Identifying the appropriate carbon contents embodied in traded products will present formidable technical difficulties, given the wide range of technologies in use around the world and very different energy resource endowments and consumption patterns among countries. In the absence of any information regarding the carbon content of the products from exporting countries, importing countries, the United States in this case, could adopt either of the two approaches to overcoming information challenges in practical implementation. One is to prescribe the tax rates for the imported product based on the domestically predominant U.S. method of production for a like product, which sets the average embedded carbon content of a particular product (Zhang 1998; Zhang and Assunção 2004). This practice is by no means without foundation. For example, the U.S. Secretary of the Treasury has adopted the approach in the tax on imported toxic chemicals under the Superfund Tax (GATT 1987; Zhang 1998). An alternative is to set the best available technology (BAT) as the reference technology level and then use the average embedded carbon content of a particular product produced with the BAT in applying border carbon adjustments (Ismer and Neuhoff 2007). Generally speaking, developing countries will bear a lower cost based on either of the approaches than using the nationwide average carbon content of imported products for the country of origin, given that less-energy-efficient technologies in developing countries produce products of higher embedded carbon contents than like products produced by more-energy-efficient technologies in the United States. However, to be more defensible, either of the approaches should allow foreign producers to challenge the carbon contents applied to their products to ensure that they will not pay for more than they have actually emitted.

CONCLUDING REMARKS

The inclusion of border carbon adjustment measures is widely considered essential to secure passage of any U.S. legislation capping its greenhouse gas emissions. Thus, on the U.S. side, in designing such trade measures, WTO rules need to be carefully scrutinized, and efforts need to be made early on to ensure that the proposed measures comply with them. After all, a conflict between the trade and climate regimes, if it breaks out, helps neither trade nor the global climate. The United States needs to explore, with its trading partners, cooperative sectoral approaches to advancing low-carbon technologies or concerted mitigation efforts in a given sector at an international level, or both. Moreover, to increase the prospects for a successful WTO defence of the Waxman-Markey type of border adjustment provision, there should be (1) a period of good faith efforts to reach agreements among the countries concerned before imposing such trade measures, (2) consideration of alternatives to trade provisions that could be reasonably expected to fulfill the same function but are not inconsistent or are less inconsistent with the relevant WTO provisions, and (3) trade provisions that can refer to the designated special international reserve allowance pool, but allow importers to submit equivalent emission reduction units that are recognized by international treaties to cover the carbon contents of imported products.

Meanwhile, being targeted by such border carbon adjustment measures, the major developing countries should make the best use of the forums provided under the UNFCCC and its Kyoto protocol to deal effectively with the proposed measures to their advantage (Zhang 2009b). The Bali Action Plan calls for "comparability of efforts" toward climate mitigation actions only among industrialized countries. However, lack of a clearly defined notion of what is comparable has led to diverse interpretations of the concept of comparability. Moreover, there is no equivalent language in the Bali Action Plan to ensure that developing country actions—whatever may have been agreed to at Copenhagen in November 2009—are comparable to those of developed countries. So, some industrialized countries, if not all, have extended the scope of its application beyond industrialized countries themselves, and are considering the term "comparable" as the standard by which to assess the efforts made by all their trading partners, in order to decide on whether to impose unilateral trade measures to address their own competitiveness concerns. Such lack of common understanding will lead each country to define whether other countries have made efforts comparable to its own. This can hardly be objective, and in turn leads one country to misuse unilateral trade measures against other trading partners to address its own competitiveness concerns.

This is not hypothetical. Rather, it is very real, as demonstrated by the Lieberman-Warner bill in the U.S. Senate and the Waxman-Markey bill in the U.S. House. If such measures became law and were implemented, trading partners might choose to challenge the United States before the WTO. A case like this is likely, given that both the top Chinese official in charge of climate issues and the Brazilian climate ambassador consider the WTO as the proper forum when developing countries are required to purchase emission allowances in the U.S. proposed cap-and-trade regime

(Samuelsohn 2007). This indicates that leading developing countries appear to be comfortable with WTO rules and institutions defending their interests in any dispute that may arise over unilateral trade measures. This is reinforced in the July 2008 Political Declaration of the Leaders of Brazil, China, India, Mexico, and South Africa (the so-called G5) in Sapporo, Japan, that "in the negotiations under the Bali Road Map, we urge the international community to focus on the core climate change issues rather than inappropriate issues like competitiveness and trade protection measures which are being dealt with in other forums."

The point, however, is that if a case like this really happens before a WTO panel, that panel would likely look to the UNFCCC for guidance on an appropriate standard for the comparability of climate efforts to assess whether that country has followed the international standard when determining comparability. Otherwise, the WTO panel will have no choice but to fall back on the aforementioned shrimp-turtle jurisprudence (see Box 6.1), and would be influenced by the fear of the political fallout from overturning U.S. unilateral trade measures in its domestic climate legislation. If the U.S. measures were allowed to stand, that would undermine the UNFCCC's legitimacy in setting and distributing climate commitments between its parties (Werksman and Houser 2008). Therefore, as strongly emphasized in my interview in the *New York Times* (Reuters 2009), there is a clear need within a climate regime to define comparable efforts toward climate mitigation and adaptation to discipline the use of unilateral trade measures at the international level, taking into account differences in their national circumstances, such as current level of development, per capita GDP, current and historical emissions, emissions intensity, and per capita emissions. If well defined, that will provide some reference for WTO panels in examining cases related to comparability issues.

Finally, it should be emphasized that the Waxman-Markey type of border adjustment provision holds out more sticks than carrots to developing countries. If the United States and other industrialized countries really want to persuade developing countries to do more to combat climate change, they should first reflect on why developing countries are unwilling to and cannot afford to go beyond the third option (Zhang 2009c) in the first place. That will require industrialized countries to seriously consider the developing countries' legitimate demand that industrialized countries need to demonstrate that they have taken the lead in reducing their own greenhouse gas emissions, to provide significant funding to support developing countries' climate change mitigation and adaptation efforts, and to transfer low- or zero-carbon emissions technologies at an affordable price to developing countries. Industrialized countries need to provide positive incentives to encourage developing countries to do more. Carrots should serve as the main means. Sticks can be incorporated, but only if they are credible and realistic and serve as a useful supplement to push developing countries to take actions or adopt policies and measures earlier than would otherwise have been the case. At a time when the world community is negotiating a post-2012 climate regime, unrealistic border carbon adjustment measures as exemplified in the Waxman-Markey bill are counterproductive in helping to reach such an agreement on comparable climate actions in the negotiations.

NOTES

1. See Zhang (1998 and 2007a), Zhang and Assunção (2004), and Charnovitz (2003) for broad discussions of potential conflicts and synergies between climate and trade regimes.
2. See Reinaud (2008) for an excellent review of practical issues involved in implementing unilateral EAR.
3. H.R. 2998, available at frwebgate.access.gpo.gov/cgi-bin/getdoc.cgi?dbname=111_ cong_bills&docid=f:h2998ih.txt.pdf.
4. This is in line with the International Brotherhood of Electrical Workers and American Electric Power proposal, which requires U.S. importers to submit allowances to cover the emissions produced during the manufacturing of those goods two years after the United States starts its cap-and-trade program (McBroom 2008).
5. To be consistent with the WTO provisions, foreign producers could arguably demand the same proportion of free allowances as U.S. domestic producers in case they are subject to border carbon adjustments.
6. In a second-best setting with pre-existing distortionary taxes, if allowances are auctioned, the revenues generated can then be used to reduce pre-existing distortionary taxes, thus generating overall efficiency gains. Parry et al. (1999), for example, show that the costs of reducing U.S. carbon emissions by 10 percent in a second-best setting with pre-existing labor taxes are five times more costly under a grandfathered carbon permits case than under an auctioned case. This is because the policy where the permits are auctioned raises revenues for the government that can be used to reduce pre-existing distortionary taxes. By contrast, in the former case, no revenue-recycling effect occurs, since no revenues are raised for the government. However, the policy produces the same tax-interaction effect as under the latter case, which tends to reduce employment and investment and thus exacerbates the distortionary effects of pre-existing taxes (Zhang 1999).
7. As part of its comprehensive strategy to control CO_2 emissions and increase energy efficiency, a carbon and energy tax had been proposed by the CEC. The CEC proposal was that member states introduce a carbon and energy tax of US$3 per barrel of oil equivalent in 1993, rising in real terms by US$1 a year to US$10 per barrel in 2000. After the year 2000 the tax rate would have remained at US$10 per barrel at 1993 prices. The tax rates were allocated across fuels, with 50 percent based on carbon content and 50 percent on energy content (Zhang 1997).
8. See Genasci (2008) for a discussion of complicating issues related to how to rebate exports under a cap-and-trade regime.
9. President Obama was quoted (Broder 2009) as saying: "At a time when the economy worldwide is still deep in recession and we've seen a significant drop in global trade, I think we have to be very careful about sending any protectionist signals out there. I think there may be other ways of doing it than with a tariff approach."

REFERENCES

Berger, J.R. 1999. Unilateral Trade Measures to Conserve the World's Living Resources: An Environmental Breakthrough for the GATT in the WTO Sea Turtle Case. *Columbia Journal of Environmental Law* 24:355–411.

Bhagwati, J., and P.C. Mavroidis. 2007. Is Action against US Exports for Failure to Sign Kyoto Protocol WTO-Legal? *World Trade Review* 6 (2): 299–310.

Bounds, A. 2006. EU Trade Chief to Reject "Green" Tax Plan. *Financial Times*, 17 December. Available at www.ft.com/cms/s/0/9dc90f34-8def-11db-ae0e-0000779e2340.html?nclick_check=1.

Bovenberg, A.L., and L.H. Goulder. 2002. Addressing Industry-Distributional Concerns in U.S. Climate Change Policy. Unpublished paper, Department of Economics, Stanford University.

Broder, J. 2009. Obama Opposes Trade Sanctions in Climate Bill. *New York Times*, 28 June. Available at www.nytimes.com/2009/06/29/us/politics/29climate.html?_r=2&scp=1&sq=oba ma%20opposes%20trade%20sanctions&st=cse.

Charnovitz, S. 2003. "Trade and Climate: Potential Conflicts and Synergies." In *Beyond Kyoto: Advancing the International Effort Against Climate Change*, edited by the Pew Center on Global Climate Change, pp. 141–70. Arlington, Virginia: Pew Center on Global Climate Change.

Economist. 2008. Pollution Law: Trading Dirt. *The Economist* (7 June): 42–44.

European Commission. 2008. *Proposal for a Directive of the European Parliament and of the Council Amending Directive 2003/87/EC so as to Improve and Extend the Greenhouse Gas Emission Allowance Trading System of the Community.* COM(2008) 16 final. Brussels: European Commission.

General Agreement on Tariffs and Trade (GATT). 1987. *United States: Taxes on Petroleum and Certain Imported Substances. Report of the Panel, Adopted on 17 June 1987.* L/6175, BISD 34S/136. World Trade Organization: Geneva. Available at www.wto.org/english/res_e/booksp_e/analytic_index_e/introduction_01_e.htm.

General Agreement on Tariffs and Trade. 1990. *Thailand: Restrictions on Importation of and Internal Taxes on Cigarettes. Report of the Panel, Adopted on 7 November 1990.* DS10/R. BISD 37S/200. Geneva: World Trade Organization. Available at www.wto.org/english/res_e/booksp_e/analytic_index_e/introduction_01_e.htm.

General Agreement on Tariffs and Trade. 1994. *United States: Restrictions on the Imports of Tuna. Report of the Panel (not adopted), circulated on 16 June 1994.* Geneva: World Trade Organization. Available at www.wto.org/english/res_e/booksp_e/analytic_index_e/introduction_01_e.htm.

Genasci, M. 2008. Border Tax Adjustments and Emissions Trading: the Implications of International Trade Law for Policy Design. *Carbon and Climate Law Review* 2 (1): 33–42.

Haverkamp, J. 2008. *International Aspects of a Climate Change Cap and Trade Program. Testimony before the Committee on Finance, U.S. Senate, February 14, 2008.* Available at http://finance.senate.gov/hearings/testimony/2008test/021408jhtest.pdf.

Houser, T. 2008. Carbon Tariffs: Why Trade Sanctions Won't Work. *China Economic Quarterly* 12 (3): 33–38.

Houser, T., R. Bradley, B. Childs, J. Werksman, and R. Heilmayr. 2008. *Leveling The Carbon Playing Field: International Competition and U.S. Climate Policy*

Design. Washington: Peterson Institute for International Economics and World Resources Institute.

Ismer, R., and K. Neuhoff. 2007. Border Tax Adjustment: A Feasible Way to Support Stringent Emission Trading. *European Journal of Law and Economics* 24 (2): 137–64.

McBroom, M. 2008. *How the IBEW-UWM-Boilermakers-AEP International Proposal Operates within Climate Legislation, June 17, 2008*. Washington: Washington International Trade Association (WITA). Available at www.wita.org/index.php?tg=fileman&idx=viewfile&idf=189&id=4&gr=Y&path=&file=WITA-+Climate+Change+-+Overview+of+IBEW-AEP+Proposal+(June+17%2C+2008).pdf.

Ministry of Commerce. 2009. *A Statement on "Carbon Tariffs," 3 July 2009.* Beijing: Ministry of Commerce of China. Available at www.mofcom.gov.cn/aarticle/ae/ag/200907/20090706375686.html (in Chinese).

Morris, M.G., and E.D. Hill. 2007. Trade Is the Key to Climate Change. *The Energy Daily* 35 (20 February): 33.

Parry, Ian W.H., R.C. Williams, III, and L.H. Goulder. 1999. When Can Carbon Abatement Policies Increase Welfare? The Fundamental Role of Distorted Factor Markets. *Journal of Environmental Economics and Management* 37 (1): 52–84.

Reinaud, J. 2008. *Issues behind Competitiveness and Carbon Leakage: Focus on Heavy Industry*. IEA Information Paper. Paris: Organisation for Economic Co-operation and Development.

Reuters. 2009. China Says "Carbon Tariffs" Proposals Breach WTO Rules. *New York Times*, July 3. Available at *www*.nytimes.com/reuters/2009/07/03/world/international-uk-china-climate.html?ref=global-home.

Samuelsohn, D. 2007. Trade Plan Opposed by China, Brazil and Mexico. *Greenwire* (26 September). Available at www.earthportal.org/news/?p=507.

Swedish National Board of Trade. 2004. *Climate and Trade Rule: Harmony or Conflict?* Stockholm: National Board of Trade.

Werksman, J., and T. Houser. 2008. *Competitiveness, Leakage and Comparability: Disciplining the Use of Trade Measures under a Post-2012 Climate Agreement*. Discussion Paper. Washington: World Resources Institute.

World Bank. 2007. *International Trade and Climate Change: Economic, Legal, and Institutional Perspectives*. Washington: World Bank.

World Trade Organization (WTO). 1998. *United States: Import Prohibition of Certain Shrimp and Shrimp Products, Report of the Appellate Body WT/DS58/AB/R*. Geneva: World Trade Organization.

World Trade Organization. 2001. *United States: Import Prohibition of Certain Shrimp and Shrimp Products, Recourse to Article 21.5 of the DSU by Malaysia, Panel Report WT/DS58/RW, Adopted on November 21, 2001*. Geneva: World Trade Organization.

Zhang, Z.X. 1997. *The Economics of Energy Policy in China: Implications for Global Climate Change*. New Horizons in Environmental Economics Series. Cheltenham, UK and Northampton, MA, USA: Edward Elgar.

Zhang, Z.X. 1998. Greenhouse Gas Emissions Trading and the World Trading System. *Journal of World Trade* 32 (5): 219–39.

Zhang, Z.X. 1999. Should the Rules of Allocating Emissions Permits be Harmonised? *Ecological Economics* 31 (1): 11–18.

Zhang, Z.X. 2000. Can China Afford to Commit Itself to an Emissions Cap? An Economic and Political Analysis. *Energy Economics* 22 (6): 587–614.

Zhang, Z.X. 2004. Open Trade with the U.S. without Compromising Canada's Ability to Comply with its Kyoto Target. *Journal of World Trade* 38 (1): 155–82.

Zhang, Z.X. 2007a. "Doing Trade and Climate Policy Together." In *Trade and Environment: A Resource Book*, edited by A. Najam, M. Halle, and R. Meléndez-Ortiz, pp. 61–62. Ottawa: International Institute for Sustainable Development; Geneva: International Center for Trade and Sustainable Development.

Zhang, Z.X. 2007b. Why Has China not Embraced a Global Cap-and-Trade Regime? *Climate Policy* 7 (2): 166–70.

Zhang, Z.X. 2009a. Multilateral Trade Measures in a Post-2012 Climate Change Regime? What Can Be Taken from the Montreal Protocol and the WTO? *Energy Policy* 37:5105–12.

Zhang, Z.X. 2009b. 美国拟征收碳关税 中国当如何应对 [How Should China Respond to the U.S. Proposed Carbon Tariffs?]. 国际石油经济 [*International Petroleum Economics*] 17 (8): 13–16. In Chinese.

Zhang, Z.X. 2009c. How Far Can Developing Country Commitments Go in an Immediate Post-2012 Climate Regime? *Energy Policy* 37 (5): 1753–57.

Zhang, Z.X., and L. Assunção. 2004. Domestic Climate Policy and the WTO. *The World Economy* 27 (3): 359–86.

7. Terms of trade in Korea: causes of decline since the mid-1990s and implications for green growth

Chin Hee Hahn and Sung-Hyun Ryu

INTRODUCTION

Korea's terms of trade have declined secularly since the mid-1990s. During the period 1996–2006, the terms of trade declined at a rate of −5.6 percent per year, reflecting both the decline in export prices and the increase in import prices. As a result, the growth rate of real gross domestic income, which was 3.0 percent per year, has not kept up with the growth rate of GDP, which was 4.6 percent. To a large extent, the gap between the production and income growth rates can explain why people's feelings about business conditions have tended to be more pessimistic than what official statistics indicate since the 1997–98 Asian financial crisis.

It is well known that the decline in Korea's terms of trade for the past decade or so has been largely driven, in an accounting sense, by the rise in import prices of crude oil and raw materials as well as the decline in export prices of semiconductor and information technology products. However, studies that explore the economic causes of the terms-of-trade decline in Korea are rare. Although the rapid growth of large developing economies, such as China and India, has been frequently pointed out to be responsible for the rise in prices of oil and raw materials, rigorous empirical studies are also hard to find.

This chapter aims to examine causes of the recent terms-of-trade decline in Korea. To set the stage, the next section evaluates the patterns of Korea's terms-of-trade decline in a broader international perspective. Specifically, this chapter decomposes the changes in terms of trade of fifty-five countries into "country price effect" and "goods price effect," following Baxter and Kouparitsas (2000), and discusses similarities and dissimilarities between Korea and other countries in their patterns of changes in terms of trade. We hope this exercise reveals some clues to the respective roles of domestic and external factors that could potentially explain the secular decline in Korea's terms of trade.

The third section employs regression analysis to examine the determinants of the terms-of-trade decline in Korea for the past decade or so. In particular, it examines whether China's trade expansion has played any role in the decline in Korea's terms of trade. With China's large and growing role in world trade as both exporter and importer, it is plausible that China's trade expansion has affected export and import prices of Korea through various channels. The focus here on China is motivated not only by the large and growing role of China in world trade but also by the existence of some external factors behind Korea's terms-of-trade decline, which is suggested by the cross-country comparison of terms-of-trade change explained above.

Many previous studies are related to this chapter, but only some of them are reviewed here. First, Baxter and Kouparitsas (2000) examine the causes of the fluctuations of terms of trade. They decompose the volatility of terms of trade of 100 countries for the period 1969–88 into goods price effect and country price effect, using data from the 1991 *World Tables* of the World Bank. Here, the goods price effect is the terms-of-trade changes due to the changes in relative prices of different goods. So, the goods price effect arises because a country's export basket differs from its import basket. For example, a country exporting manufactures and importing oil will experience a terms-of-trade decline when the price of oil relative to manufactures rises. Meanwhile, the country price effect is the changes in terms of trade that are not the goods price effect. This effect basically arises from the deviation from the law of one price with different prices for the same good across countries. In the Baxter and Kouparitsas study, the country price effect captures the terms-of-trade changes induced by the changes in export price relative to the import price of the same good (for example manufactures).[1]

In presenting the results, they first classify countries into developed and developing countries, and then further classify them into three subgroups: commodity exporters, fuel exporters, and exporters of manufactures. Their results suggest that the country price effect has a non-negligible role in explaining terms-of-trade fluctuations. For developing-country manufactures exporters, in particular, it is shown that most of the terms-of-trade volatility is attributable to the country price effect, although most of the terms-of-trade fluctuations of fuel exporters is explained by the goods price effect. While the decomposition methodology of this study is basically taken from their study, this chapter focuses on medium- to long-term changes, rather than short-term fluctuations, in terms of trade. Also, this chapter uses highly disaggregated trade data (UN Comtrade

SITC 5 digit, Rev. 3) to construct export and import price indexes for each country, while their study uses fairly aggregated data from the 1991 *World Tables* of the World Bank. Given that the decomposition results could be sensitive to the level of aggregation being used to measure each category of "goods," the advantage of using disaggregated data is that it allows us to examine the sensitivity of the results in relation to the level of aggregation.

Next, it is hard to find existing studies that examine the impact of China's trade expansion on Korea's terms-of-trade change.[2] Kim (2006), the only exception that we are aware of, shows that China's recent trade expansion had a negative effect on terms of trade of Mexico.[3] In estimating the impact, Kim (2006) does not use Mexico's export and import prices but uses averages of export and import price indexes of the United States as proxies for the international prices. So, in his paper, the effect of China's trade expansion on Mexico's terms of trade is indirectly estimated through its impact on international prices. This chapter differs from Kim (2006) in several respects. One difference is that this chapter uses Korea's unit value of exports and imports calculated at the SITC five-digit level to estimate the impact. We think this is a more direct approach. Another difference is that this chapter separately estimates the impact of China's trade expansion on export prices and import prices of Korea. This approach does not rule out the existence of the country price effect in terms-of-trade changes.[4]

Finally, this chapter is related to the literature on the Prebisch-Singer hypothesis. Prebisch (1950) and Singer (1950) argued that the terms of trade of developing-country commodity exporters are likely to exhibit a securely declining trend due to a low income elasticity of demand for primary products, a low entry barrier, an infinitely elastic labor supply condition, and so on. Controversies in the subsequent literature were about whether terms of trade of developing countries secularly declined, rather than about which factors determine long-run changes of terms of trade of developing countries. Recently, Kaplinski (2006) has argued that the terms of trade of developing countries have declined secularly since the Second World War and that low innovation intensity, as well as low entry barriers of exports, of developing countries were the prime causes. Kaplinski goes on further to argue that low innovation intensity and low entry barriers of exports are not product-specific but country-specific characteristics. We hope this chapter sheds some light on these issues by documenting terms-of-trade changes of various country groups and decomposing them into goods price and country price effects.

DECOMPOSITION OF TERMS-OF-TRADE CHANGES OF KOREA AND INTERNATIONAL COMPARISON

Recent Trends of Terms of Trade in Korea

As mentioned in the introduction, Korea's terms of trade have secularly declined since the mid-1990s, reflecting both a decline in export prices and an increase in import prices (Figure 7.1 and Table 7.1). The pattern of the terms-of-trade decline, however, changed around the turn of the century. In the second half of the 1990s, the decline in terms of trade (−6.5 percent per year) was driven by the decline in export prices (−9.6 percent), which was larger than the decline in import prices (−3.1 percent). In contrast, the terms-of-trade decline in the 2000s has been driven by a rise in import prices. In accounting, the decline in export prices in the second half of the 1990s was driven by semiconductors and information technology products, while the increase in import prices in the 2000s has been driven by products such as oil, raw materials, and grain.

Was the terms-of-trade decline for the past decade or so a phenomenon that is Korea-specific? Or was it common across a group of countries? What are the respective roles of goods and country price effects in its decline? How do they compare with other countries' experiences? The following analysis addresses these questions.

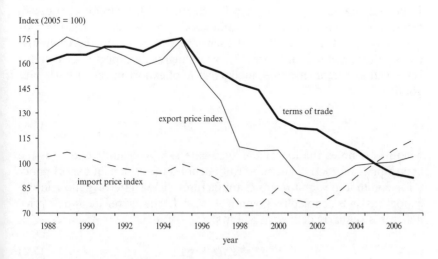

Figure 7.1   Indexes of export and import prices and change in terms of trade in Korea, 1988–2007

Table 7.1   Korea's export and import price indexes and growth of terms of trade, 1988–2007 (%)

Item	1988–95	1995–2000	2000–07	1995–2007
Terms of trade	1.2	−6.5	−4.7	−5.5
Export price index	0.6	−9.6	−0.5	−4.3
Semiconductors	−20.7	−40.7	−28.4	−33.5
Precision equipment	1.2	−3.7	−8.3	−6.4
Information and communication equipment	2.0	−12.4	−1.8	−6.2
Electric machines for domestic purposes	3.2	−6.3	−3.5	−4.6
Import price index	−0.6	−3.1	4.2	1.1
Electric machines for domestic purposes	0.2	−9.0	−7.5	−8.1
Information and communication equipment	−1.9	−8.2	−7.6	−7.8
Precision equipment	2.3	−3.3	−4.9	−4.2
Semiconductors	1.2	4.0	−5.0	−1.2
Crude oil	1.6	8.9	12.7	11.1

Source: Bank of Korea.

Decomposition Methodology

Following Baxter and Kouparitsas (2000), we decompose changes in terms of trade into country price and goods price effects as follows. For convenience, let us suppose that there are two goods in the economy: commodities and manufactures. Then the aggregate export price deflator is defined as a geometric weighted average of export prices of particular goods

$$p^x = s_c^x p_c^x + s_m^x p_m^x \qquad (7.1)$$

where p^x denotes the log of the aggregate export deflator and p_i^x is the export price of good i. The subscript i denotes the particular export good: c for commodities and m for manufactures. The share of good i in the export basket is denoted by s_i^x and $s_c^x + s_m^x = 1$. The aggregate import price deflator is similarly defined as follows.

$$p^m = s_c^m p_c^m + s_m^m p_m^m \qquad (7.2)$$

Then, the log of terms of trade is defined as follows.

$$\ln TOT = p^x - p^m \tag{7.3}$$

After some manipulation, equation (7.3) can be decomposed into the country price effect and the goods price effect. Here, the country price effect captures the effect arising from different prices for the same good. So, the country price effect involves the term $(p_c^x - p_c^m)$, for example. The goods price effect involves the term denoting relative prices of different goods: $(p_c^x - p_m^x)$, for example.

There is, however, no unique way of decomposing terms of trade into goods price and country price effects. As explained by Baxter and Kouparitsas (2000), there are two sets of decompositions depending on whether one uses the export share or the import share as the weight in the calculation of the country price effect. In the case where the export share is used as the weight, the terms of trade can be decomposed as follows:

$$\ln TOT = p^x - p^y = \underbrace{s_c^x(p_c^x - p_c^m) + s_m^x(p_m^x - p_m^m)}_{\text{country price effect}} + \underbrace{(s_c^x - s_c^m)(p_c^m - p_m^m)}_{\text{goods price effect}} \tag{7.4}$$

When the import share is used as the weight, then the terms of trade can be decomposed alternatively as follows:

$$\ln TOT = p^x - p^y = \underbrace{s_c^m(p_c^x - p_c^m) + s_m^m(p_m^x - p_m^m)}_{\text{country price effect}} + \underbrace{(s_c^x - s_c^m)(p_c^x - p_m^x)}_{\text{goods price effect}} \tag{7.5}$$

In equations (7.4) and (7.5), the goods price effect term is expressed with manufactures being used as the numeraire.

More generally, when there are n goods in the economy, there are $2n$ ways of decomposing terms of trade into country price effect and goods price effect. That is, there are two ways of decomposition, depending on whether one uses the export share or the import share as the weight in expressing the country price effect and, for each of them, there are n ways of decomposition depending on the choice of the numeraire good. It should be noted, however, that the choice of numeraire good does not affect the country price terms. This means that it does not affect the magnitude of the total goods price effect. The choice of numeraire good affects only the way that the goods price effect is further decomposed into subcomponents. So, inasmuch as breaking down terms of trade into the two effects is concerned, there are effectively two distinctive ways of decomposition.

The decomposition of terms of trade in the n-good case is briefly explained below. As before, the log of the terms of trade is defined as follows:

$$\ln TOT = p^x - p^m = \sum_i s_i^x p_i^x - \sum_i s_i^m p_i^m \qquad (7.6)$$

When the export share is used as the weight in expressing the country price effect, the terms of trade can be decomposed as (M.1: methodology 1)

$$\ln TOT = \sum_i s_i^x (p_i^x - p_i^m) + \sum_{i \neq k} (s_i^x - s_i^m)(p_i^m - p_k^m) \qquad (7.7)$$

where the subscript k denotes the numeraire good. When the import share is used as the weight in expressing country price effect, the decomposition of terms of trade can be expressed as follows (M.2: methodology 2).

$$\ln TOT = \sum_i s_i^m (p_i^x - p_i^m) + \sum_{i \neq k} (s_i^x - s_i^m)(p_i^x - p_k^x) \qquad (7.8)$$

Thus, when equation (7.7) is used, the change in terms of trade is decomposed as

$$\Delta \ln TOT = \Delta \left(\sum_i s_i^x (p_i^x - p_i^m) \right) + \Delta \left(\sum_{i \neq k} (s_i^x - s_i^m)(p_i^m - p_k^m) \right) \quad (7.9)$$

where Δ denotes the change of the relevant term from year t to $t + \tau$.

Export and Import Prices of Goods

In the previous subsection, we explained the methodology of decomposing changes in terms of trade when export and import prices of particular goods are available. In this subsection, we explain how export and import prices of particular goods are constructed. Aggregate export and import price deflators for a number of countries can be constructed based on the data on value and quantity of exports and imports from the *World Development Indicators* (WDI) of the World Bank. For the purposes of this chapter, however, we need export and import prices of particular goods based on common methodology and data across countries, which are not readily available. Thus, for each country where relevant data are available, we calculated the export and import prices of particular goods using the five-digit (SITC Rev. 3) product level UN Comtrade data. In this study, we assume that there are three goods in an economy: commodities

(SITC 0, 1, 2, 4, and 68), fuel (SITC 3), and manufactures (SITC 5 and 6 excluding 68), as in Baxter and Kouparitsas (2000).

In constructing export and import price indexes of a particular good, we eliminated some products with extreme unit values from the basket in two steps. In the first step, we eliminated those products whose year-on-year growth rate of unit value belongs to the upper or the lower 5 percentile of its distribution. Even after this procedure, there were products remaining whose unit value of export (or import) increased or decreased unrealistically. So, in the second step, we additionally eliminated those products with unit value of export (or import) larger (smaller) than its previous year's value by a factor of more than one hundred. As a consequence, in the case of year 2005, the remaining products account for about 88 percent of the total in terms of product coverage, 99 percent of the total export value, and 98 percent of the total import value.

The products that are eliminated in the second step are spread over twelve countries and explained in Table 7.2. One alternative way of eliminating products with extreme unit values of exports or imports might be to apply a criterion of 10 percentile rather than 5 percentile. However, since the extreme observations tend to be concentrated on a small number of countries, two-step procedures were taken in this chapter to minimize the loss of observations.

The export or import price index of a particular good—manufactures, for example—was then constructed as a chained-Paache index following the Bank of Korea, based on unit values of exports or imports at product level. The procedure can be briefly explained as follows. We first define this year's export (or import) price relative to last year's, which we call the ratio of export prices between adjacent years, as follows.

$$\frac{\sum_i PQ}{\sum_i P_{-1}Q}$$

Then, a particular year's export price index is obtained by multiplying all ratios of export prices between adjacent years from the base year to the comparison year with the export price index in the base year, which is equal to 100.[5]

The countries in our sample were determined as follows. Among the 200 or so countries covered in the UN Comtrade data, countries for which both export and import price indexes can be calculated for each good for all years from 1996 to 2006 were included in our "full sample," which comprises 68 countries.

However, there is no guarantee that the behaviors of the aggregate

Table 7.2   Excepted products in twelve selected countries by SITC number

Country and year	Com-modities	Fuel (SITC Rev. 3)	Manu-factures
Bolivia			
2002 import		3449 (1)	
2002 export		32121, 32222, 3223, 3250, 3330, 33542 (6)	
Canada			
2003 import		3211, 32122, 3221, 32221, 3250, 3330, 33542 (7)	
2003 export	11	32121, 32222, 3223, 3250, 33542 (5)	
2003 import	23	3211, 32121, 32122, 3221, 32221, 3250, 3432 (7)	
Columbia			
2000 import	55	32121, 3223, 3330, 33419, 33429, 33541, 33542, 33543 (8)	110
India			
1999 import		3221, 32122, 3223, 3250 (4)	
2000 export		3221, 32122, 3223, 3250, 33521 (5)	
2000 import		32121, 32122, 3221, 32221, 3250 (5)	
Korea			
1993 export		33411, 33421, 3343, 3344 (4)	
Mexico			
2001 import			122
Malaysia			
1999 import		32221, 32222, 3223, 3250, 33411, 33429, 3344, 3345, 33512, 33531, 33532, 33541, 33542, 33543 (14)	
2000 import		32221, 3223, 33411, 33429, 3344, 3345, 33512, 33523, 33531, 33532, 33541, 33542, 33543 (13)	
Oman			
2001 export		3330 (1)	
Philippines			
2002 export		32222 (1)	
Slovak Republic			
2002 export		3441 (1)	
2004 export		3211, 3510 (2)	
Slovenia			
2004 export		33521 (1)	

Notes: We eliminated products whose year-on-year growth rates of unit value belong to the upper and lower five percentiles of their distribution. Even after this procedure, we additionally eliminated products with unit values of export (or import) larger (or smaller) than the previous year's value by more than one hundred times.

export and import price deflators of a country calculated in this chapter will coincide with those provided by the official statistics, since the data and methodology underlying the official aggregate export and import price deflators might well vary across countries. So, it is worthwhile to compare aggregate export and import price deflators calculated in this chapter with official statistics. Considering the aggregate export and import price deflators calculated from the WDI data set as the official statistics, Table 7.3 shows the correlations of the two annual growth rates of the aggregate export (or import) deflators. Depending on the patterns of correlations, the 68 countries were classified into four subgroups.

The first group (group A) comprises 20 countries for which the aggregate export and import price deflators can be calculated from the WDI data set, and for which correlations of both export and import price deflators are positive and significant at the 10 percent level. The second group (group B) consists of 17 countries for which either an export or an import price deflator shows a positive and significant correlation at the 10 percent level, but the other aggregate price deflator does not exhibit any significant correlation. The third group (group C) is the 13 countries for which neither correlation is significant at the 10 percent level. The final and fourth group (group D) is 18 countries for which the aggregate export or import price deflator cannot be calculated from the WDI data set. These are the countries for which it is impossible to examine the correlations. The basic sample of countries analyzed in this chapter comprises groups A, B, and D, for a total of 55 countries.[6]

In the case of Korea, the aggregate export and import price deflators published by the Bank of Korea are available, so that it is possible to compare them with those calculated in this study and with those calculated from the WDI data set. Figures 7.2 and 7.3 show the terms of trade, the export price deflator, and the import price deflator from the three sources, in relation to the level of the price index and to the growth rates. Above all, the movements of the aggregate export and import price deflators from the WDI data set almost exactly match those published by the Bank of Korea. The aggregate export and import price deflators calculated in this chapter seem to capture fairly well the short- and long-run characteristics of the deflators of the Bank of Korea, although the differences in their levels tends to become larger in years that are further away from the base year (2005). For example, the decline in export price deflators during the second half of the 1990s as well as the increase of the import price deflator since the beginning of the 2000s is visible in both series. As a result, the terms-of-trade decline since the mid-1990s is also visible in both series.[7] Also, the annual growth rates of export and import price deflators calculated in this study closely mimic those of the officially published deflators.

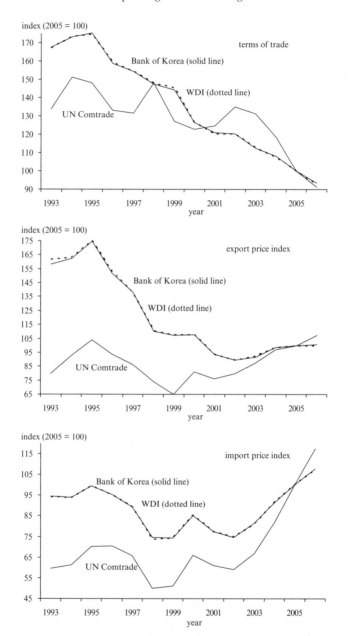

Figure 7.2   *Terms of trade and export and import prices in Korea:*
           comparison of Bank of Korea, World Development Index, and
           UN Comtrade indexes, 1993–2006

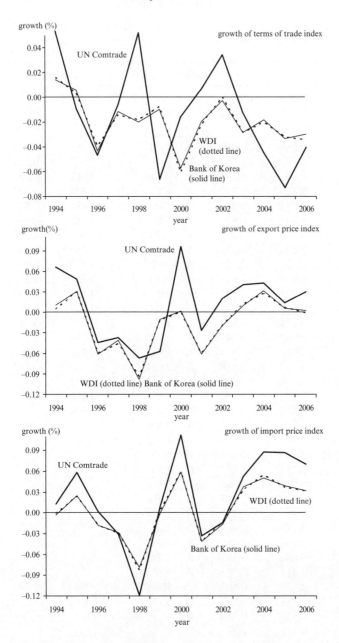

Figure 7.3   Growth of terms of trade and export and import price indexes in Korea, 1994–2006

Table 7.3   *Correlations of growth rates of export and import price indexes in 68 countries*

Country	Export price index	Import price index	Group	Country	Export price index	Import price index	Group
ARG	0.986	0.909	A	COL	0.953	−0.128	B
BRA	0.977	0.664	A	HUN	0.967	−0.207	B
ESP	0.863	0.785	A	VEN	−0.012	0.726	B
HKG	0.821	0.739	A	CIV	0.325	0.402	C
IDN	0.623	0.915	A	CRI	0.481	0.527	C
ITA	0.991	0.825	A	FRA	0.455	−0.030	C
JPN	0.955	0.982	A	GRC	−0.903	0.844	C
KEN	0.650	0.673	A	HND	0.150	−0.090	C
KOR	0.709	0.991	A	MYS	0.322	0.037	C
NOR	0.986	0.933	A	NLD	0.255	−0.246	C
NZL	0.903	0.801	A	OMN	0.431	0.832	C
PAK	0.645	0.927	A	PHL	0.477	0.134	C
PER	0.934	0.914	A	PRT	0.575	−0.113	C
POL	0.917	0.847	A	TUN	0.356	−0.121	C
SLV	0.674	0.635	A	IND	−0.571	−0.265	C
SWE	0.822	0.658	A	MEX	−0.440	−0.143	C
THA	0.629	0.754	A	JAM	0.282		D
TUR	0.832	0.854	A	LVA	0.852		D
URY	0.876	0.859	A	AUT			D
USA	0.968	0.807	A	BRB			D
AUS	0.768	0.528	B	CHE			D
CAN	0.889	0.608	B	CZE			D
CHN	0.234	0.883	B	EST			D
DEU	0.926	0.986	B	FIN			D
DNK	0.469	0.607	B	GBR			D
DZA	0.971	0.337	B	HRV			D
ECU	0.920	0.102	B	ISL			D
GTM	0.904	0.256	B	KAZ			D
IRL	0.686	0.588	B	LTU			D
MAR	0.412	0.782	B	MKD			D
NER	0.339	0.602	B	ROM			D
NIC	0.782	0.617	B	SVK			D
SGP	0.342	0.820	B	SVN			D
BOL	0.716	−0.131	B	TTO			D

Note: Depending on the patterns of correlations between the WDI dataset and the UNComtrade data set, the sixty-eight countries were classified into four subgroups.

*Table 7.4 Pairwise correlations of export and import price indexes and
terms of trade: Bank of Korea, World Development Index, and
the present study*

	Bank of Korea	WDI	Present study
Terms of trade			
Bank of Korea	1	0.991	0.477
WDI		1	0.474
Present study			1
Export price index			
Bank of Korea	1	0.997	0.781
WDI		1	0.776
Present study			1
Import price index			
Bank of Korea	1	0.997	0.986
WDI		1	0.986
Present study			1

Note: Annual growth correlations.

Table 7.4 shows that the correlations between the two series are 0.781 for
the export price deflator and 0.986 for the import price deflator. In the case
of the terms of trade, however, the correlation coefficient is low at 0.477,
so that one needs to be cautious about analyzing the short-run behavior of
the terms of trade calculated in this study.[8]

Trade Structure and Changes in Terms of Trade

Before discussing the results from the terms-of-trade decomposition, we
briefly examine the relationship between trade structure of a country and
medium- to long-term changes in terms of trade, which is in Table 7.5
for our 55-country sample during the period from 1996 to 2006. Table
7.5 shows, for each country subgroup, the growth rate of terms of trade
as well as the average share of commodities, fuel, and manufactures in
total exports and imports. The net exports share is the export share minus
import share. The 55 countries were classified into developing and developed countries, which were further divided into commodity exporters, fuel
exporters, and exporters of manufactures depending on the goods with
the largest net export share. The figures in the table are the averages of the
corresponding country groups with the period-average exports as weights.
Table 7.6 shows similar results for the period 1993–2006, but with the
number of countries reduced to 40.

Table 7.5 Change of terms of trade and trade structure: 55 selected countries, 1996–2006

	No. of countries	Growth (%)		Terms of trade	Export shares (%)			Import shares (%)			Net export shares (%)		
		Export price index	Import price index		Commodities	Fuel	Manufactures	Commodities	Fuel	Manufactures	Commodities	Fuel	Manufactures
Developing countries													
Commodities	21	0.8	1.6	-0.8	36.1	4.2	59.8	13.8	10.2	76.0	22.2	-6.0	-16.2
Fuel	6	4.4	-1.9	6.3	21.5	41.2	37.3	20.7	8.4	71.0	0.9	32.8	-33.7
Manufactures	10	2.3	5.3	-2.9	15.3	3.9	80.8	19.1	11.8	69.1	-3.8	-7.9	11.7
Total	37	2.2	3.6	-1.4	21.1	7.9	71.1	18.0	11.0	71.0	3.1	-3.2	0.1
Developed countries													
Commodities	7	1.5	4.3	-2.8	36.6	7.0	56.4	22.4	13.0	64.6	14.2	-6.0	-8.2
Fuel	2	6.5	3.9	2.5	11.3	16.5	72.2	15.5	3.8	80.8	-4.2	12.8	-8.6
Manufactures	9	1.4	5.1	-3.7	9.2	0.6	90.2	18.4	10.1	71.6	-9.2	-9.4	18.6
Total	18	2.1	4.8	-2.7	16.1	4.2	79.7	19.0	10.0	71.1	-2.9	-5.8	8.7
World													
Commodities	28	1.3	3.5	-2.2	36.4	6.1	57.5	19.7	12.1	68.2	16.7	-6.0	-10.7
Fuel	8	5.9	2.4	3.5	14.0	23.0	62.9	16.8	5.0	78.2	-2.8	18.1	-15.3
Manufactures	19	1.7	5.1	-3.4	11.1	1.7	87.3	18.6	10.7	70.8	-7.5	-9.0	16.5

Table 7.6 Change of terms of trade and trade structure: 40 selected countries, 1993–2006

	No. of countries	Growth (%)			Export shares (%)			Import shares (%)			Net export shares (%)		
		Export price index	Import price index	Terms of trade	Commodities	Fuel	Manufactures	Commodities	Fuel	Manufactures	Commodities	Fuel	Manufactures
Developing countries													
Commodities	12	1.5	2.1	−0.7	34.9	4.3	60.8	13.3	10.2	76.5	21.5	−5.9	−15.6
Fuel	5	3.9	−1.6	5.4	20.5	40.3	39.3	21.6	6.6	71.8	−1.2	33.7	−32.5
Manufactures	5	3.0	5.0	−2.0	17.1	4.3	78.5	20.5	10.8	68.7	−3.4	−6.4	9.8
Total	22	2.7	3.5	−0.8	21.1	7.8	71.2	19.2	10.3	70.6	1.9	−2.5	0.6
Developed countries													
Commodities	7	2.3	4.3	−2.0	37.1	6.4	56.4	22.8	14.1	63.0	14.3	−7.7	−6.6
Fuel	2	5.9	4.8	1.2	11.5	18.3	70.2	15.5	3.7	80.8	−4.0	14.6	−10.6
Manufactures	9	2.0	4.8	−2.8	8.8	0.6	90.5	18.7	10.5	70.8	−9.8	−9.9	19.7
Total	18	2.6	4.7	−2.1	18.2	4.4	77.4	19.7	10.9	69.4	−1.5	−6.5	8.0
World													
Commodities	19	2.1	3.7	−1.6	36.6	5.9	57.5	20.5	13.1	66.4	16.1	−7.2	−8.9
Fuel	7	5.4	3.2	2.2	14.4	25.2	60.4	17.4	4.6	77.9	−3.1	20.6	−17.5
Manufactures	14	2.3	4.9	−2.6	12.1	2.1	85.8	19.4	10.6	70.0	−7.3	−8.5	15.8

The first column of Table 7.5 shows the number of countries in each country group. In the case of developing countries, a majority (21 out of 37) were classified as commodity exporters, while 10 countries were classified as exporters of manufactures and the remaining 6 as fuel exporters. In contrast, 9 out of 18 developed countries were classified as exporters of manufactures, while 7 countries were commodity exporters and the remaining 2 were fuel exporters. Overall, the table shows that the comparative advantage in manufacturing is more pronounced in developed countries than in developing countries, which might suggest that the role of manufacturing in Korea's growth in the future cannot be disregarded.[9]

On average, the terms of trade of developing countries declined at an annual rate of −1.4 percent and those of developed countries at −2.7 percent, during the period 1996–2006. What is noticeable is that the terms of trade of fuel exporters as a group increased at an annual average of 3.5 percent, regardless of the level of development. By contrast, there were declines in the terms of trade of exporters of manufactures (−3.4 percent) as well as commodity exporters (−2.2 percent). In this regard, there is no noticeable difference between developing and developed countries, except that the terms-of-trade decline in exporters of manufactures is more pronounced for developed countries and that the terms-of-trade improvement of fuel exporters is more pronounced for developing countries.

As discussed above, the medium- to long-term trend in a country's terms of trade for the period analyzed here tends to be driven mainly by the goods that give the country a comparative advantage, not by whether the country is a developing or developed country. Fuel exporters experienced a terms-of-trade improvement, and exporters of manufactures experienced a terms-of-trade deterioration, regardless of their level of income. In this regard, the experience since the mid-1990s is clearly at odds with the arguments along the lines of Prebisch-Singer, which emphasize country specific factors as long-term determinants of terms of trade.

Decomposition of Changes in Terms of Trade

Results for Korea
Table 7.7 shows the results from the decomposition of terms-of-trade change in Korea based on equations (7.7) and (7.8), which use exports and imports shares, respectively, as weights in country price terms. We first discuss the contributions from the country price effect and the goods price effect, and then continue with a more detailed breakdown.

The terms of trade, calculated in this study, declined at an annual average of −4.5 percent for the period 1993–2006. What should be noted is that the decline in the terms of trade in Korea was driven by the goods

Table 7.7   Decomposition of change in Korea's terms of trade, 1993–2006
(%)

Export shares weighted

Period	Export price index	Import price index	Terms of trade	Goods price	Country price	Goods price effect		Country price effect		
						A1	A2	A3	A4	A5
1993–96	1.7	6.0	−4.3	−4.4	0.1	−1.4	−3.0	−0.3	−0.1	0.5
1996–01	−4.1	−1.5	−2.6	−3.7	1.1	0.2	−4.0	0.0	0.1	1.0
2001–06	6.7	13.4	−6.7	−10.4	3.8	−1.8	−8.7	0.0	−0.1	3.9
1996–2006	1.3	6.0	−4.6	−7.1	2.5	−0.8	−6.3	0.0	0.0	2.4
1993–2006	1.4	6.0	−4.6	−6.5	1.9	−0.9	−5.6	−0.1	0.0	2.0

Import shares weighted

Period	Export price index	Import price index	Terms of trade	Goods price	Country price	Goods price effect		Country price effect		
						B1	B2	B3	B4	B5
1993–96	1.7	6.0	−4.3	0.8	−5.1	0.3	0.5	−1.7	−3.3	−0.2
1996–2001	−4.1	−1.5	−2.6	−5.3	2.7	0.5	−5.8	0.1	2.4	0.2
2001–06	6.7	13.4	−6.7	−4.7	−2.0	−0.7	−4.0	−0.2	−2.2	0.4
1996–2006	1.3	6.0	−4.6	−5.0	0.4	−0.1	−4.9	0.0	0.1	0.3
1993–2006	1.4	6.0	−4.6	−3.7	−0.9	0.0	−3.7	−0.4	−0.7	0.2

Notes: 1. Goods price is A1+A2, B1+B2. Country price is A3+A4+A5, B3+B4+B5.
c (commodity), f (fuel), m (manufactures). 2. A1= (axc-amc)*(pmc-pmm), A2= (axf-amf)*(pmf-pmm), A3= axc*(pxc-pmc), A4= axf*(pxf-pmf), A5= axm*(pxm-pmm), B1= (axc-amc)*(pxc-pxm), B2= (axf-amf)*(pxf-pxm), B3= amc*(pxc-pmc), B4= amf*(pxf-pmf), B5= amm*(pxm-pmm). 3. aij is individual items' export (import) shares. i=x, m. x (export), m (import). pij is individual items' export (import) price index. j=c, f, m. c (commodity), f (fuel), m (manufactures).

price effect. The contribution from the goods price effect was −6.5 percentage point for M.1, and −3.6 percentage point for M.2. The role of the country price effect differs between M.1 and M.2. In the case of M.1, the country price effect is about 1.9 percentage points, which played the role of offsetting the negative effect from the goods price effect. In the case of M.2, however, the country price effect is small but negative.

Turning to the results for subperiods, we find that for the period 1993–96 the relative importance of country and goods price effects in explaining the terms-of-trade decline is different between the two methodologies. In the case of M.1, all of the decline in the terms of trade is accounted for by the goods price effect, whereas it is driven by the country price effect in the case of M.2. In the periods after 1996, the terms-of-trade decline is mostly driven by the goods price effect regardless of periods or methodologies. During the period 1996–2006, the terms of trade in Korea declined at an annual average of −4.6 percent. Of this decline, the goods price effect accounts for −7.1 percentage point in M.1 and −5.0 percentage point in M.2. That is, the goods price effect alone more than explains away the actual terms-of-trade decline during that period. The role of the country price effect differs between the two methodologies: it is somewhat large (2.5 percentage points) in M.1, but small (0.4 percentage point) in M.2.

Then, what relative price changes have driven the large goods price effect? The columns on the right hand side of Table 7.7 show a detailed breakdown of goods price and country price effects. The large and negative goods price effect is driven by the rise in the price of fuel relative to manufactures, or the fall in the price of manufactures relative to fuel, in both M.1 and M.2. This result is consistent with our earlier discussion that Korea's terms-of-trade decline was driven mainly by the fall in the prices of semiconductors and information technology products during the second half of the 1990s and by the rise in oil prices in the 2000s. What this decomposition exercise shows is that the observed terms-of-trade decline reflects changes in relative prices of different goods, not the changes in export prices relative to import prices of the same good.

The country price effect, if any, is accounted for mostly by manufactures. The rise in export prices of manufactures relative to import prices accounts for almost all of the total country price effect during the period from 1996 to 2006, regardless of methodologies. Furthermore, the country price effect arising from manufactures became larger in the 2000s than in the second half of the 1990s. The contribution of manufactures to the country price effect is larger in M.1 than in M.2, reflecting the larger share of manufactures in exports than in imports. Overall, the rise in export prices of manufactures relative to import prices has contributed to offsetting the decline in overall terms of trade, especially in the 2000s.

Results for the 55-country sample

We now evaluate Korea's terms-of-trade decline since the mid-1990s from a broad international perspective. To do so, we discuss the results of the decomposition for the 55-country sample. Tables 7.8 and 7.9 are the results when the country price effect is weighted by the export share and the import share, respectively.

For exporters of manufactures—whether developing or developed countries, and including Korea—the goods price effect accounts for a substantial portion of the decline in the terms of trade, regardless of methodologies. In the case of M.1, the goods price effect accounts for most of the decline in terms of trade, while in the case of M.2, it explains about half of

Table 7.8   Decomposition of change in terms of trade for 55 countries: export share weighted, 1996–2006 (%)

	Coun-tries (no.)	Export price index	Import price index	Terms of trade	Goods price	Coun-try price	Goods price A1	A2	Country price A3	A4	A5
Developing countries											
Commodities	21	0.8	1.6	−0.8	−0.6	−0.2	1.0	−1.5	−0.4	0.0	0.2
Fuel	6	4.4	−1.9	6.3	−2.5	8.9	−1.3	−1.3	0.6	8.3	0.0
Manufactures	10	2.3	5.3	−2.9	−1.9	−1.0	0.0	−2.0	−0.5	−0.8	0.3
Total	37	2.2	3.6	−1.4	−1.6	0.2	0.1	−1.8	−0.4	0.4	0.2
Developed countries											
Commodities	7	1.5	4.3	−2.8	−1.2	−1.6	0.5	−1.7	−0.6	−1.0	0.1
Fuel	2	6.5	3.9	2.5	2.0	0.5	−0.4	2.4	0.1	−0.1	0.5
Manufactures	9	1.4	5.1	−3.7	−4.0	0.3	0.0	−4.0	0.1	−0.3	0.5
Total	18	2.1	4.8	−2.7	−2.6	−0.1	0.1	−2.6	−0.1	−0.4	0.4
World											
Commodities	28	1.3	3.5	−2.2	−1.0	−1.1	0.6	−1.6	−0.5	−0.7	0.1
Fuel	8	5.9	2.4	3.5	0.8	2.7	−0.6	1.5	0.2	2.1	0.4
Manufactures	19	1.7	5.1	−3.4	−3.3	−0.1	0.0	−3.4	−0.1	−0.4	0.4

Notes: 1. Goods price is A1+A2. Country price is A3+A4+A5. c (commodity), f (fuel), m (manufactures). 2. A1= (axc-amc)*(pmc-pmm), A2= (axf-amf)*(pmf-pmm), A3= axc*(pxc-pmc), A4= axf*(pxf-pmf), A5= axm*(pxm-pmm). 3. aij is individual items' export (import) shares. i=x, m. x (export), m (import). pij is individual items' export (import) price index. j=c, f, m. c (commodity), f (fuel), m (manufactures). 4. Entries refer to export-weighted average of the group (Baxter and Kouparitsas 2002).

the decline in terms of trade of developing and developed countries. What is noticeable here is that the goods price effect mainly arises from the rise of prices of fuel relative to manufactures (or decline in the prices of manufactures relative to fuel).

For fuel exporters, it is hard to find any systematic pattern from the decomposition results. For commodity exporters, the terms-of-trade decline in developing and developed countries tends to be driven by the country price effect. We will not discuss the results for these groups of countries further, since the main focus of this study is on comparing Korea with other exporters of manufactures.

The decomposition results for exporters of manufactures are fairly similar to those obtained for Korea, which we discussed earlier. Specifically, these countries as a group experienced a fairly large decline in their terms of trade for the period 1996–2006, and a large portion of the decline is accounted for by the goods price effect that stems from the relative price changes between manufactures and fuel.

Table 7.9 Decomposition of change in terms of trade for 55 countries: import share weighted, 1996–2006 (%)

	Coun-tries (no.)	Export price index	Import price index	Terms of trade	Goods price	Coun-try price	Goods price B1	B2	Country price B3	B4	B5
Developing countries											
Commodity	21	0.8	1.6	−0.8	1.0	−1.8	0.1	0.9	−0.1	−1.8	0.1
Fuel	6	4.4	−1.9	6.3	3.4	2.9	−1.5	4.9	1.5	0.8	0.7
Manufactures	10	2.3	5.3	−2.9	−0.9	−2.0	0.4	−1.3	−0.9	−0.9	−0.3
Total	37	2.2	3.6	−1.4	0.0	−1.4	0.1	−0.1	−0.4	−0.9	−0.1
Developed countries											
Commodity	7	1.5	4.3	−2.8	0.2	−3.0	−0.2	0.4	−0.2	−2.7	0.0
Fuel	2	6.5	3.9	2.5	2.7	−0.1	−0.4	3.1	0.2	−0.9	0.6
Manufactures	9	1.4	5.1	−3.7	−2.4	−1.2	0.0	−2.4	0.1	−1.4	0.1
Total	18	2.1	4.8	−2.7	−1.2	−1.5	−0.1	−1.1	0.0	−1.7	0.1
World											
Commodity	28	1.3	3.5	−2.2	0.5	−2.6	−0.1	0.5	−0.2	−2.4	0.0
Fuel	8	5.9	2.4	3.5	2.9	0.7	−0.7	3.6	0.5	−0.4	0.6
Manufactures	19	1.7	5.1	−3.4	−2.0	−1.5	0.1	−2.1	−0.2	−1.3	0.0

Notes: 1. Goods price is B1+B2. Country price is B3+B4+B5. c (commodity), f (fuel), m (manufactures). 2. B1= (axc-amc)*(pxc-pxm), B2= (axf-amf)*(pxf-pxm), B3= amc*(pxc-pmc), B4= amf*(pxf-pmf), B5= amm*(pxm-pmm). 3. aij is individual items' export (import) shares. i=x, m. x (export), m (import). pij is individual items' export (import) price index. j=c, f, m. c (commodity), f (fuel), m (manufactures). 4. Entries refer to export-weighted average of the group (Baxter and Kouparitsas 2002).

This finding has an important implication concerning the ultimate causes of Korea's terms-of-trade decline since the mid-1990s. That is, Korea's decline was probably caused mainly not by Korea-specific factors but by some factors that commonly affected exporters of manufactures across the world. We think China's rise and rapid trade expansion is clearly one possible candidate.

THE IMPACT OF CHINA'S TRADE EXPANSION ON KOREA'S TERMS-OF-TRADE DECLINE

During the past several decades, China has exhibited remarkably high and sustained growth by following an export-led development strategy. Consequently, China has recently become one of the largest economies in the world in terms of GDP and trade. Also, there are studies documenting that China's export structure is more sophisticated than can be predicted by the level of income and that its exports are rapidly changing from labor-intensive to capital- and skill-intensive products (see for example Lall and Albaladejo 2004; Rodrik 2006; Schott 2006; and Kim et al. 2006).

Given the large size of the economy and the volume of its trade, it is likely that China has had a nonnegligible influence on the prices of a wide range of products. Indeed, it has frequently been suggested in the mass media that China is responsible for the decline in the relative prices of manufactures (thereby contributing to the low inflation of the major developed countries) and for the rise in the prices of oil and raw materials. Given this background, we examine below whether and how China's trade expansion affected Korea's terms-of-trade decline, based on regression analysis.

Methodology and Data

The expansion of exports from China is expected to decrease the corresponding international prices by increasing the supply of those products in the world market. Similarly, the expansion of China's imports is expected to raise the corresponding international prices by increasing the demand for those products in the world market. Hence, inasmuch as Korea's export and import prices are related to the international prices of corresponding goods, China's export expansion is expected to lower Korea's export prices and China's import expansion is expected to raise Korea's import prices of corresponding products. Nevertheless, China's trade expansion can have differential effects on export and import prices of Korea. Suppose that Korea's imports in the fuel category are concentrated on crude oil and that Korea's exports in the same category are mostly refined oil products. Suppose also that China's imports in the fuel category are mostly crude oil as in Korea. Under these circumstances, the expansion of China's imports in the fuel category will directly affect the fuel import prices of Korea, but will only indirectly affect the fuel export prices of Korea. That is, if differences exist in a more detailed product composition in exports and imports within a broadly defined product category, then China's trade expansion can have differential effects on export and import prices of Korea.

In order to analyze the effects of China's trade expansion on Korea's export and import prices, we set up the regression equation as follows:

$$DP_i^j = c + \beta \cdot SXC_i + \gamma SMC_i + \delta Z_i + \varepsilon_i \qquad (7.10)$$

where the dependent variable DP_i^j is the growth rate of the export ($j = X$) or import ($j = M$) price of product i, SXC_i is China's world export market share (or its change), and SMC_i is China's world import market share (or its change), in product i. Z_i denotes the vector of control variables. The above equations were estimated using both product level trade data and industry level data. In the case of industry level data, we considered the productivity growth rate, R&D intensity, capital intensity, and skill intensity as control variables.

As discussed already, the coefficients on β are expected to be negative while the coefficients on γ are expected to be positive, in both export price and import price regressions. Meanwhile, the coefficient on the productivity growth rate is expected to be negative in both regressions, since faster productivity growth implies a faster outward shift of the market supply curve.[10] We use R&D intensity, capital intensity, and skill intensity to control for possible effects of these variables on the export and import prices.

The regression equation was estimated using both product and industry level data. The product level data comes from SITC five-digit (Rev. 3) level UN Comtrade data. The industry data were constructed by linking the trade data with KSIC five-digit industry data from the Mining and Manufacturing Census.[11] In the case of product level data, the control variables cannot be constructed because of lack of information. However, there is an advantage in using product level data, because all products that are included in the trade data can be used for the analysis.

We ran separate regressions for the three periods 1993–96, 1996–2001, and 2001–06. As discussed earlier, the terms of trade in Korea have secularly declined since the mid-1990s, but its patterns are different between the second and third subperiods. The dependent variable is annual average growth of unit value of exports or imports. China's export market share is measured either as period-averages (MSXC) or as changes from the beginning to the end of the period (DSXC). China's import market share, MSMC or DSMC, is similarly measured. The productivity growth rate of an industry is measured as the growth rate of value added per employee. R&D intensity is R&D expenditure divided by production. Capital intensity is measured as the tangible fixed assets per employee. Skill intensity is measured as the share of non-production workers in employment. R&D intensity, capital intensity, and skill intensity are the averages of the corresponding period.

Regression Results

Results from product level trade data
Table 7.10 shows the regression results using product level trade data. As expected, in regressions of both export price growth and import price

Table 7.10 Regression of China's rise effects on Korea: product-level trade data, 1993–2006

Dependent variable	Model 1			Model 2			Model 3		
	1993–96	1996–2001	2001–06	1993–96	1996–2001	2001–06	1993–96	1996–2001	2001–06
Export price index									
DSXC	−0.040	−0.240**	−0.488***				0.008	−0.256**	−0.332**
	(0.190)	(0.117)	(0.162)				(0.202)	(0.122)	(0.169)
DSMC	−0.376*	−0.153	0.148				−0.461**	−0.131	0.077
	(0.213)	(0.158)	(0.208)				(0.219)	(0.164)	(0.213)
MSXC				−0.022	−0.005	−0.132***	−0.030	0.006	−0.113***
				(0.041)	(0.026)	(0.034)	(0.044)	(0.027)	(0.035)
MSMC				0.074	−0.016	0.053	0.108	−0.022	0.051
				(0.069)	(0.041)	(0.052)	(0.071)	(0.042)	(0.053)
R^2	0.001	0.002	0.004	−0.000	−0.001	0.007	0.001	0.001	0.008
Observations (no.)	2021	2091	2141	2021	2091	2141	2021	2091	2141
Import price index									
DSXC	−0.221	−0.639***	−0.476***				−0.226	−0.630***	−0.400***
	(0.140)	(0.096)	(0.120)				(0.147)	(0.099)	(0.124)
DSMC	0.105	0.053	0.527***				0.077	0.039	0.378**
	(0.169)	(0.139)	(0.160)				(0.171)	(0.145)	(0.168)
MSXC				−0.014	−0.032*	−0.078***	0.002	−0.005	−0.056**
				(0.031)	(0.019)	(0.024)	(0.032)	(0.020)	(0.025)
MSMC				0.072	0.047	0.121***	0.069	0.011	0.094**
				(0.054)	(0.037)	(0.040)	(0.055)	(0.038)	(0.042)
R^2	0.000	0.018	0.010	−0.000	0.001	0.009	0.000	0.017	0.015
Observations (no.)	2316	2348	2371	2316	2348	2371	2316	2348	2371

Notes: Heteroskedasticity-consistent standard errors are in parentheses. Coefficients with asterisks are 1 (***), 5 (**), and 10 (*) percent levels, respectively.

growth, the change in China's world export market share (DSXC) is estimated with a negative and significant coefficient in most specifications
for the two subperiods 1996–2001 and 2001–06. For the period 2001–06,
MSXC is also estimated with a negative and significant coefficient in
both export price and import price growth regressions. Although neither
China's world import market share (MSMC) nor its change (DSMC)[12]
is estimated with a significant coefficient in export price growth regressions, it is estimated with a significantly positive coefficient in import price
growth regressions, especially in the period 2001–06.[13]

We discuss in some detail the regression results for the period 2001–06,
when China's influence on the world economy is considered to have become
stronger. In export price growth regressions, when we measure China's
export expansion as DSXC or MSXC, each of them is estimated with a
significantly negative coefficient (models 1 and 2). When we include both
variables as explanatory variables in model 3, they are both significantly
negative. By contrast, we could not find any significantly positive effect of
China's import expansion on the export price growth of Korea. Thus, it
is suggested that, at least since the beginning of the 2000s, China's trade
expansion had an effect of lowering the growth of export prices of Korea.

In import price growth regressions for the period 2001–06, China's
export expansion significantly lowered the growth of import prices of
Korea. This result is consistent with the casual observation that a wide
range of manufactured products became available at cheaper prices due
to China's export expansion. However, the estimation results suggest
that the expansion of China's imports significantly raised the growth of
Korea's import prices, especially in the 2000s. Given the high import share
of oil and raw materials in both Korea and China, this result might not be
surprising.

Based on the regression results of Table 7.10, it is possible to make a
rough quantitative assessment of the effect of China's trade expansion
on Korea's export and import price growth and, hence, on the change in
terms of trade. The result of this exercise, based on model 3 for the period
2001–06, are shown in Figures 7.4 and 7.5, demonstrating the effects on
Korea's export prices and import prices, respectively. Each figure displays
the effect on the aggregate export or import price deflator, as well as the
effects on six disaggregated goods.[14] For each good, the first bar shows
the effect of China's export expansion on the export (or import) price
growth of the corresponding good, while the second bar shows the effect
of China's import expansion. Here, the size of the bar is the average estimated effect for the products that belong to the corresponding goods, that
are weighted by the period-average export value or import value.[15] The
third bar shows the net effect.

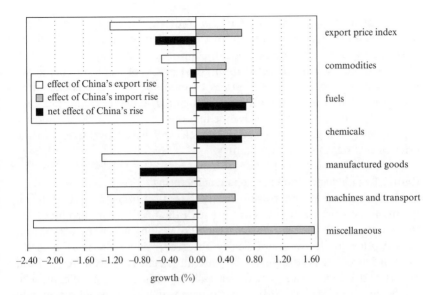

Figure 7.4   Effects of China's rise on the growth of Korea's export price index (model 3), 2001–06

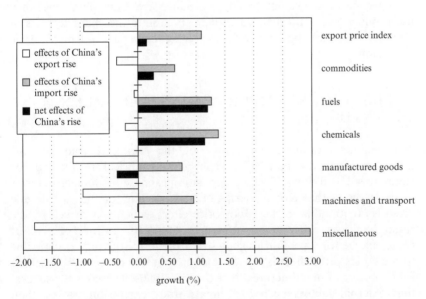

Figure 7.5   Effects of China's rise on the growth of Korea's import price index (model 3), 2001–06

As shown, in the aggregate, China's trade expansion appears to have contributed to the decline in terms of trade in Korea during the 2000s by lowering export prices and raising import prices. At a disaggregated level, China's trade expansion raised the import prices of fuel and chemicals, in particular, and lowered the prices of manufactured goods excluding chemicals. This result is consistent with our discussion in the previous section that the terms of trade decline in Korea during the 2000s is mostly explained by the goods price effect, which was driven by the rise in the price of fuel relative to manufactures.

Results from industry level data on manufacturing
Table 7.11 shows the regression results based on industry level data on the manufacturing sector. In contrast with the results from product level trade data, the effects from China's trade expansion on Korea's export and import price growth were not clearly observed.[16] Nevertheless, the coefficient of DSXC was estimated to be generally negative as expected in both export and import price growth regressions, although insignificant. The coefficient of DSMC was estimated to be positive as expected but mostly insignificant in export price growth regressions. Meanwhile, no coherent sign patterns were observed on the coefficient of DSMC in import price growth regressions. In interpreting the results, however, we prefer to put more weight on the results based on product level trade data, since they use product level data detailed enough to capture the characteristics of markets which are appropriate for analyzing the impact of China's trade expansion.

SUMMARY AND IMPLICATIONS FOR GREEN GROWTH

This chapter examines the causes of the terms of trade decline in Korea since the mid-1990s. The main empirical results can be summarized as follows. The decomposition exercise of changes in terms of trade shows that Korea's terms of trade decline for the past decade or so is attributable to the goods price effect, which was driven by the rise of oil prices relative to manufactures. The decomposition of terms of trade change for 55 countries shows that the terms-of-trade decline due to the goods price effect is a phenomenon that was commonly observed for exporters of manufactures since the mid-1990s. These results suggest that external factors such as China's trade expansion, rather than internal factors, are mostly responsible for the decline in the terms of trade.

Table 7.11   *Regression of China's rise effects on Korea: industry-level manufacturing data, 1993–2006*

Dependent variable	Model 1		Model 2		Model 3	
	1996–2000	2000–03	1996–2000	2000–03	1996–2000	2000–03
Export price index						
DSXC	0.343	−0.279	0.254	−0.237	0.266	−0.230
	(0.613)	(0.191)	(0.618)	(0.203)	(0.620)	(0.199)
DSMC	0.025	0.688*	0.264	0.546	0.259	0.510
	(0.947)	(0.368)	(0.971)	(0.378)	(0.973)	(0.370)
R&D intensity			−0.003	0.013*	−0.003	0.010
			(0.002)	(0.008)	(0.002)	(0.008)
Capital intensity			−0.008	0.002	−0.013	0.000
			(0.008)	(0.008)	(0.009)	(0.009)
Productivity growth					0.112*	−0.103*
					(0.063)	(0.060)
Technology intensity					0.000	0.001
					(0.001)	(0.001)
R^2	−0.005	0.010	−0.003	0.011	0.002	0.019
Observations (number)	365	363	364	360	359	352
Import price index						
DSXC	−0.422	−0.129	−0.292	−0.094	−0.213	−0.123
	(0.526)	(0.168)	(0.527)	(0.176)	(0.529)	(0.170)
DSMC	1.010***	−0.078	0.900**	−0.011	0.870**	−0.061
	(0.430)	(0.333)	(0.429)	(0.345)	(0.431)	(0.333)
R&D intensity			0.001	0.014*	0.002	0.014*
			(0.002)	(0.007)	(0.002)	(0.007)
Capital intensity			0.017**	0.002	0.016**	−0.003
			(0.007)	(0.008)	(0.007)	(0.008)
Productivity growth					−0.012	0.009
					(0.054)	(0.054)
Technology intensity					0.001*	0.000
					(0.001)	(0.001)
R^2	0.011	−0.004	0.026	0.002	0.031	−0.001
Observations (number)	369	371	368	367	363	357

Notes: Heteroskedasticity-consistent standard errors are in parentheses. Coefficients with asterisks are 1 (***), 5 (**), and 10 (*) percent levels, respectively.

In accordance with these results, the regressions of export and import price growth suggest that China's trade expansion contributed to Korea's terms of trade decline. Specifically, China's export expansion is estimated to have lowered the growth of both export and import prices of Korea. In contrast, China's import expansion is estimated to have raised the growth of import prices, particularly in the 2000s, while its effect on the growth of export prices is not clearly observable. In sum, the empirical evidence of this study suggests that China's trade expansion contributed to the decline in the terms of trade of Korea, especially in the 2000s, by raising the import prices of oil and raw materials and lowering the export prices of manufactured products.

The empirical evidence of this study suggests that Korea's terms of trade decline might persist if China's economic growth and trade expansion continue in the future. This possibility cannot be excluded, especially if Korea maintains an economic structure that is highly dependent on imported energy. Thus, Korea needs to have an appropriate policy framework in place in order to reduce the high energy dependency of the economy, reduce the high external dependency of energy, and promote the development of related technologies. In this respect, the empirical evidence of this study provides some support for the related policies in the "green growth" strategies of Korea.

Furthermore, in order to alleviate the pressure on Korea's export prices arising from China's rapid growth as well as the increasing sophistication of its exports, it is necessary for Korea to maintain an export basket that is differentiated from China's. Continuous upgrading of the quality of human capital as well as the innovation capability of Korea's economy seems to be essential in this regard.

NOTES

1. Baxter and Kouparitsas (2000) explain that the country price effect can arise from producers' pricing-to-market behavior, which means that producers price the same good differently across countries where it is sold. Meanwhile, Goldberg and Knetter (1996) explain that the deviations from the law of one price can arise from pricing-to-market behavior as well as incomplete exchange rate pass-through. They discuss a broad range of existing empirical studies.
2. We will not discuss in detail here many existing studies of China's impact on the economies of Korea and other developing countries. See, for example, Li (2002), Lall and Albaladejo (2004), World Bank (2006), Choi et al. (2005), Hahn and Choi (2007), Ahn, Fukao, and Ito (2011), Kim et al. (2006), Ito and Hahn (2010), Hahn (2006) and literature cited therein.
3. Kim (2006) also shows some evidence that China's trade expansion had a negative effect on terms of trade of Korea as well, although most of the discussion is devoted to the impact on Mexico.

4. Kim (2006) is effectively assuming away the country price effect in terms of trade changes, by constructing one international price for one product. In other words, his analysis is based on the assumption of the law of one price.

5. For a more detailed explanation, see the *Explanations of the Economic Indicators Made Easy*, issued by the Bank of Korea.

6. The results discussed further in this section do not change qualitatively even if group D countries are excluded from the sample.

7. The mild increase in the level of export price deflator calculated in this chapter, however, is somewhat in contrast with the roughly stagnant behavior of the official export price deflator. In fact, it is rather natural to observe some discrepancies between the two deflators, given the differences in the data and methodology. The Bank of Korea uses HSK 10-digit trade data to construct the aggregate export and import price deflators, but eliminates a large number of observations as outliers. So, in the case of the year 2005, the products included in the calculation cover only 11.4 percent of the total number of products, 78.4 percent of the total export value, and 79.4 percent of the total import value. As explained before, a much larger share of products is used in this study.

8. This problem is not of much concern in this study, because we are focusing on medium- or long-run behavior of terms of trade.

9. That is, although the share of services sector in Korea will increase further as the level of income rises, this should not be understood as implying that the role of manufacturing in Korea's growth will decline with the rise of income.

10. When the correlations of the industry level productivity growth rate among countries are high, a negative coefficient can be expected.

11. We used data on manufacturing only, by excluding the mining sector.

12. The variable DSMC was estimated with a significantly negative coefficient in the period 1993–96, contrary to our expectation. We could not find any satisfactory explanation for this result.

13. When we estimated the regressions in Table 7.10 with fixed effect models with year dummies, we could obtain qualitatively similar results: China's export expansion lowered the growth of both export and import prices of Korea, while China's import expansion raised the growth of import prices only. The results are available from the authors upon request.

14. The SITC five-digit products were first classified as before into three goods (commodities, fuel, and manufactures) and then the manufactures were further disaggregated into four subcategories.

15. The calculated effect in this way should not be considered as the "effect of China's trade expansion" in a rigorous sense. In fact, a meaningful "counterfactual" is needed to discuss the effect of China's trade expansion. However, it is not at all clear which counterfactual is meaningful. The counterfactual implicitly assumed in the above exercise is the situation where the relevant explanatory variables are zero, which is certainly very unrealistic. In this respect, this exercise should be considered as one convenient way of interpreting the regression results in a quantitative matter.

16. Various factors are responsible for the differences in results between product level trade data and industry level data, such as the level of aggregation, time period analyzed, and coverage of the data. However, industry coverage is not likely to be one of those factors, since we were able to obtain qualitatively similar results when we used the product level data but confined its coverage to manufacturing products.

REFERENCES

Ahn, Sanghoon, Kyoji Fukao, and Keiko Ito. 2011. "Globalization and Labor Markets in East Asia." In *Reforms for Korea's Sustained Growth*, edited by

Chin Hee Hahn and Sang-Hyop Lee, pp. 157–97. Seoul: Korea Development Institute.

Baxter, Marianne, and Michael A. Kouparitsas. 2000. *What Causes Fluctuations in the Terms of Trade?* Working Paper 7462. Cambridge, Massachusetts: National Bureau of Economic Research.

Choi, Yong-Seok, Moon Joong Tcha, and Jong-il Kim. 2005. *China's Growing Economy and Its Implications for the Korean Economy.* Research Monograph 2005-04. Seoul: Korea Development Institute. In Korean.

Goldberg, Pinelopi, and Michael Knetter. 1996. Measuring the Intensity of Competition in Export Market. Mimeographed paper.

Hahn, Chin Hee. 2006. *Import Competition from China and OECD Countries and Patterns of Plant Growth in Korean Manufacturing.* Seoul: Korea Development Institute. In Korean.

Hahn, Chin-Hee, and Yong-Seok Choi. 2007. "The Impact of China on Output and Investment Growth in Korean Manufacturing." In *Economic Growth of Korea after the Financial Crisis: Evaluation and Implications,* edited by Inseok Shin and Chin Hee Hahn. Research Monograph 2007-05. Seoul: Korea Development Institute. In Korean.

Ito, Takatoshi, and Chin Hee Hahn, eds. 2010. *The Rise of China and Structural Changes in Korea and Asia.* Cheltenham, UK, and Northampton, MA, USA: Edward Elgar.

Kaplinsky, Raphael. 2006. Revisiting the Revisited Terms of Trade: Will China Make a Difference? *World Development* 34 (6): 981–95.

Kim, Chong-Sup. 2006. The Effect of China's Trade Expansion on Mexican Exports. Mimeographed paper.

Kim, Joon-Kyung, Yangseon Kim, and Chung H. Lee. 2006. *Trade, Investment and Economic Integration of South Korea and China.* Working Paper 2006-01. Seoul: Korea Development Institute.

Lall, Sanjaya, and Manuel Albaladejo. 2004. China's Competitive Performance: A Threat to East Asian Manufactured Exports? *World Development* 32 (9): 1441–66.

Li, Y. 2002. *China's Accession to WTO: Exaggerated Fears.* Discussion Paper 165. Geneva: United Nations Conference on Trade and Development.

Prebisch, R. 1950. *The Economic Development of Latin America and Its Principal Problems.* Economic Bulletin for Latin America 7. New York: United Nations.

Rodrik, Dani. 2006. *What's So Special about China's Export?* Working Paper 11947. Cambridge, Massachusetts: National Bureau of Economic Research.

Schott, Peter K. 2006. *The Relative Sophistication of Chinese Exports.* Working Paper 12173. Cambridge, Massachusetts: National Bureau of Economic Research.

Singer, H.W. 1950. The Distribution of Gains between Investing and Borrowing Countries. *American Economic Review* 15:478–85.

World Bank. 2006. *An East Asian Renaissance: Ideas for Economic Growth.* Washington: World Bank.

8. Low carbon green growth and energy policy in Korea

Jin-Gyu Oh

INTRODUCTION

The Intergovernmental Panel on Climate Change (IPCC), which was awarded the 2007 Nobel Peace Prize, provided key findings in its fourth assessment report (Pachauri and Reisinger 2007). The findings include the fact that eleven of the warmest years, since instrumental records have been kept, occurred during the previous twelve years, along with additional evidence for the acceleration of climate change. The IPCC also reported that the past 100-year increase was 0.74 degrees Celsius (°C). If we do not take immediate action to curb greenhouse gas (GHG) emissions, the average temperature is projected to increase in the range of 1.1° to 6.4° C by the end of this century. In addition, 20–30 percent of plant and animal species would be in danger of extinction if the temperature increase were to exceed 1.5–2.5° C. These findings suggest the imperativeness of taking action to mitigate climate change.

Under the 1997 Kyoto protocol, developed countries are bound to reduce their GHG emissions by an average of 5.2 percent below 1990 levels between 2008 and 2012, and the international community has initiated negotiations for the post-2012 period. These negotiations address the commitment of developed countries and developing countries after 2012. Korea assumes the obligation as a developing country under the Kyoto protocol. Ranked tenth in terms of GHG emissions, Korea needs to be proactive in mitigating climate change through aggressive national policies and measures.

The energy sector accounts for 84 percent of GHG emissions in Korea, both posing challenges and offering opportunities for the transition to a low carbon economy. Furthermore, response strategies for low carbon challenges should translate into green growth opportunities for Korea. This chapter reviews the current status of Korea's GHG inventory, energy consumption, and carbon dioxide (CO_2) emissions. It also provides the most up-to-date projections of CO_2 emissions up to 2030 and evaluates the main driving forces for such projections. In addition, energy policies for addressing low carbon green growth strategies are introduced in depth. Finally, an overall assessment of energy policies is provided.

CURRENT STATUS

Korea's National GHG Inventory

This section discusses the GHG inventory of Korea, categorized by sector and by specific gas type. The total "gross" emissions of GHGs, consisting of six gases (CO_2, CH_4, N_2O, HFCs, PFCs, and SF_6), increased to 599.5 million tons of CO_2 equivalent (tCO_2eq) in 2006, from a baseline 298.1 million tCO_2eq in 1990, resulting in average annual increases of 4.5 percent for the entire period (Table 8.1). This indicates an exact doubling of GHG emissions for the sixteen-year period from 1990 to 2006 (the latest year for which the GHG inventory is available). Major sources of emissions include the energy, industrial process, agriculture, and wastes sectors. Forestry and land uses act as carbon sinks, sequestering CO_2 from the atmosphere.

In 2006, the energy sector was the single most significant emitter, accounting for approximately 84 percent of total GHG emissions. The share of the industrial process sector was about 11 percent, while the wastes and agriculture sectors recorded shares of 2.6 percent and 2.5 percent, respectively. The uptake of GHGs through forestry and land-use changes was estimated at 31.2 million tCO_2eq for the same year, accounting for about 5.2 percent of total gross emissions. Therefore, total "net" emissions reached 568.4 million tCO_2eq (see Table 8.1).

Observing GHG emissions by gas type reveals that the component share of CO_2 was largest at 88.8 percent, in 2006, followed by CH_4 (4.2 percent) and N_2O (2.6 percent) (see Table 8.2).

Energy-related CO2 Emissions

Emissions of CO_2 from the combustion of fossil fuels (energy-related CO_2) amounted to 497.1 million tCO_2eq in 2006 (see Table 8.3), accounting for 83 percent of the total gross emissions. Since these energy-related CO_2 emissions are the most important factor in the climate change debate, they will be discussed in more detail later in this chapter.

In 2006, CO_2 emissions from fossil fuel combustion, or energy-related CO_2, reached 497.1 million tons for an average annual increase of 4.7

Table 8.1 *Greenhouse gas emissions and removals by sector in Korea,*
1990–2006

Sector	1990	2000	2005	2006	Average annual growth rate 1990–2006 (%)
Energy					
Emissions (mtCO$_2$eq)	247.7	438.5	498.5	505.4	4.6
Share of gross (%)	83.1	82.6	83.9	84.3	
Industrial process					
Emissions (mtCO$_2$eq)	19.9	58.3	64.8	63.7	7.5
Share of gross (%)	6.7	11.0	10.9	10.6	
Agriculture					
Emissions (mtCO$_2$eq)	13.5	17.0	16.1	15.1	0.7
Share of gross (%)	4.5	3.2	2.7	2.5	
Waste management					
Emissions (mtCO$_2$eq)	17.0	17.2	14.9	15.4	−0.6
Share of gross (%)	5.7	3.2	2.5	2.6	
Gross emissions					
Emissions (mtCO$_2$eq)	298.1	531.0	594.4	599.5	4.5
Share of gross (%)	100.0	100.0	100.0	100.0	
Forest and land use change (mtCO$_2$eq)	−23.7	−37.2	−32.0	−3.12	1.7
Net emissions (mtCO$_2$eq)	274.4	493.8	562.4	568.4	4.7

Note: mtCO$_2$eq = million tons of carbon dioxide equivalent.

Source: KEEI (2008).

percent since 1990. Per capita CO_2 emissions from fossil fuel combustion increased from 5.57 tons in 1990 to 10.29 tons in 2006. Carbon intensity, defined as CO_2 per unit of energy, has steadily decreased at an average annual rate of 1.2 percent since 1990.

The electricity generation sector accounted for the largest share with 36.1 percent of total energy-related CO_2 emissions in 2006 (Table 8.4). Energy-related CO_2 emissions from the industrial sector accounted for 31.7 percent, the transportation sector for 20 percent, the residential and commercial sector for 11.4 percent, and the public sector for 0.9 percent. Owing to the expansion of coal-fired power plants, the electricity sector showed the steepest increase in CO_2 emissions, with average annual growth of more than 10 percent from 1990 to 2006. Emissions from the transportation sector similarly showed a very high growth rate (5.5 percent).

Table 8.2   Emissions of major gases in Korea, 1990–2006

Type of gas	1990	2000	2005	2006	Average annual growth rate 1990–2006 (%)
Carbon dioxide (CO_2)					
Emissions (mt CO_2eq)	257.7	466.3	525.1	532.2	4.6
Share of gross (%)	86.4	87.8	88.3	88.8	
Methane (CH_4)					
Emissions (mt CO_2eq)	36.6	28.0	25.2	25.3	−2.3
Share of gross (%)	12.3	5.3	4.2	4.2	
Nitrous oxide (N_2O)					
Emissions (mt CO_2eq)	2.9	14.4	18.0	15.5	11.1
Share of gross (%)	1.0	2.7	3.0	2.6	
Total					
Emissions (mt CO_2eq)	298.1	531.0	594.4	599.5	4.5
Share of total (%)	100.0	100.0	100.0	100.0	

Notes: Hydrofluorocarbons (HFCs), perfluorocarbons (PFCs), and sulfur hexafluoride (SF_6) are not shown in this table since they are negligible.
Data do not necessarily add up to the respective column totals, because of rounding errors.

Source: KEEI (2008).

Table 8.3   Energy-related CO_2 emissions in Korea: main indicators 1990–2006

Sector	1990	2000	2005	2006	Average annual growth rate 1990–2006 (%)
CO_2 (million tons) (A)	239.0	432.2	490.5	497.1	4.7
Population (thousands)	42 869	47 008	48 138	48 297	0.7
Energy (thousand toe) (B)	93 192	192 887	228 622	233 372	5.9
Per capita CO_2 emissions (tons)	5.57	9.19	10.19	10.29	3.9
Carbon intensity (tons/toe) (A/B)	2.56	2.24	2.15	2.13	−1.2

Note: toe = tons of oil equivalent.

Source: KEEI (2008).

Table 8.4   Energy-related CO_2 emissions by sector in Korea, 1990–2006

Sector	1990	2000	2005	2006	Average annual growth rate 1990–2006 (%)
Electric power					
Emissions (mt CO_2eq)	37.9	125.7	170.8	179.3	10.2
Share of total (%)	15.9	29.1	34.8	36.1	
Industrial					
Emissions (mt CO_2eq)	87.2	152.4	156.2	157.5	3.8
Share of total (%)	36.5	35.3	31.8	31.7	
Transportation					
Emissions (mt CO_2eq)	42.2	86.6	97.5	99.3	5.5
Share of total (%)	17.7	20.0	19.9	20.0	
Residential and commercial					
Emissions (mt CO_2eq)	64.7	63.5	61.1	56.7	−0.8
Share of total (%)	27.1	14.7	12.5	11.4	
Public					
Emissions (mt CO_2eq)	7.0	4.0	4.9	4.3	−3.0
Share of total (%)	2.9	0.9	1.0	0.9	
Total					
Emissions (mt CO_2eq)	239.0	432.2	490.5	497.1	4.7
Share of total (%)	100.0	100.0	100.0	100.0	

Source: KEEI (2008).

Main Assumptions in Projections of Energy Demand and CO_2 Emissions to 2030

As shown in the previous section, CO_2 emissions from energy consumption are the single most important factor in GHG emissions in Korea. Therefore, a long-term picture of the energy sector is important in developing strategies for GHG abatement. This section presents the long-term energy outlook and CO_2 emissions for Korea up to 2030.

Various forces drive energy demand. Gross domestic product (GDP), industrial structure, population, and the quantity of energy-consuming products (such as vehicles and appliances) are especially noteworthy. In general, there is a big difference between developed countries and developing countries in terms of GDP and population. For the projection of CO_2 emissions in Korea, GDP is assumed to maintain strong growth at an annual rate of 4.7 percent from 2005 to 2010. GDP is projected to increase 4.0 percent from 2010 to 2020 and 3.5 percent from 2020 to

2030, for an average annual rate of 3.0 percent throughout the entire period. Population is expected to rise slightly from 48.29 million in 2005 to 49.96 million in 2020, and decline to 49.33 million in 2030, resulting in an annual increase rate of only 0.1 percent throughout the entire period.

The share of three energy-intensive manufacturing industries (petrochemicals, nonmetallic minerals, and primary metals) of total manufacturing is expected to decline for the entire period, from 26.6 percent in 2005 to 19.5 percent in 2020, and 17.3 percent in 2030.

The most important factor that determines transportation energy demand is the number of vehicles being driven. Vehicle ownership will continue to grow rapidly as personal income increases. For example, the number of passenger cars is expected to increase very rapidly, from 10.8 million in 2005 to 19.5 million in 2020, and 20.1 million in 2030, for an average annual increase of 2.5 percent.

The number of households is a major driving force in residential energy demand, but it is assumed to increase at a low annual average of only 0.6 percent throughout the same period. The number of persons per household is assumed to decrease from 3.06 in 2005 to 2.68 in 2030.

Floor space in buildings is a driving factor for commercial energy demand and is assumed to increase at a high annual average of 5.4 percent throughout the same period.

Energy Projections to 2030

As shown in Table 8.5, demand for primary energy is projected to grow at a slower rate than GDP through 2030. Total primary energy demand is expected to increase at an annual rate of 2.7 percent from 2005 to 2010, 2.3 percent from 2010 to 2020, and 2.0 percent from 2020 to 2030. The average for the entire period is 2.3 percent.

Total primary energy consumption is projected to increase from 229.3 million tons of oil equivalent (toe) to 400.1 million toe between 2005 and 2030, for an average annual increase of 2.3 percent (Table 8.6). During the same period, total final energy consumption is projected to increase from 172.1 million toe in 2005 to 289.9 million toe in 2030, for an average annual increase of 2.1 percent.

Annual GDP growth rates are forecast to be 4.7 percent from 2005 to 2010, 4.0 percent from 2010 to 2020, and 3.5 percent from 2020 to 2030—producing an average annual rate of 3.9 percent. Energy and GDP elasticities for these same periods are expected to be 0.57, 0.58, and 0.57 (or an average of 0.57 percent). The energy to GDP elasticity ratio was high at 0.91 for the period 1981–90 and very high at 1.38 for the period 1990–97,

Table 8.5   Major energy and economic indicators in Korea, 2005–30

	2005	2010	2020	2030	Average annual growth rate (%)			
					2005– 10	2010– 20	2020– 30	2005– 30
Primary energy (mtoe)	229.3	261.5	328.8	400.1	2.7	2.3	2.0	2.3
Per capita energy (toe/person)	4.75	5.31	6.58	8.11	2.3	2.2	2.1	2.2
Population (million)	48.3	49.2	50.0	49.3	0.4	0.1	−0.1	0.1
Energy/GDP (toe/million won)	0.32	0.29	0.24	0.21	−2.0	−1.6	−1.5	−1.7
Energy-GDP elasticity (%)					0.57	0.58	0.57	0.57

Note: Data for 2010 are estimates. mtoe = million tons of oil equivalent.

Source: KEEI (2006).

but then dropped to 0.81 for the period 1998–2005, after the foreign currency crisis hit Korea in 1998.

This suggests that the foreign currency crisis was a turning point that weakened the relationship between GDP and energy use from as high as 1.38 during the 1990s to 0.57 for the period 2005–10. Note, however, that the energy and GDP elasticity is expected to remain constant at around 0.57 from the year 2005 to 2030. This poses a key challenge to Korea: to decouple energy growth from GDP growth.

Energy intensity, measured in terms of primary energy consumption per unit of GDP, is projected to decline by 1.7 percent per year through the period up to 2030, reflecting the improved efficiency of energy usage. Energy use per capita is expected to increase by 2.2 percent per year, reaching 8.11 toe in 2030.

As shown in Table 8.6, petroleum use dominates the energy mix in Korea. It accounted for 44 percent of primary energy demand in 2005 and will continue to account for the largest share through 2030. Liquefied natural gas (LNG) is the fastest growing fuel source, increasing at a projected average rate of 3.9 percent per year during 2005–30. In fact, it is projected to account for 20 percent of the total primary energy supply in 2030. Demand for natural gas is projected to increase by a factor of more than 2.5, from 30.0 million toe in 2005 to 78.6 million toe in 2030. Nuclear power, which has zero CO_2 emissions, is expected to double from 36.7

Table 8.6   Primary energy demand in Korea by source, 2005–30

Energy source	2005	2010	2020	2030	Average annual growth rate (%)			
					2005–10	2010–20	2020–30	2005–30
Coal								
mtoe	54.8	63.8	72.7	82.1	3.1	1.3	1.2	1.6
Share of total demand (%)	23.9	24.4	22.1	20.5				
Petroleum								
mtoe	101.6	109.0	123.5	142.0	1.4	1.3	1.4	1.3
Share of total demand (%)	44.3	41.7	37.6	35.5				
Liquefied natural gas								
mtoe	30.0	38.0	60.0	78.6	4.9	4.7	2.7	3.9
Share of total demand (%)	13.1	14.5	18.3	19.6				
Hydropower								
mtoe	1.3	1.4	1.4	1.5	1.2	0.5	0.3	0.6
Share of total demand (%)	0.6	0.5	0.4	0.4				
Nuclear power								
mtoe	36.7	41.7	56.8	74.2	2.6	3.1	2.7	2.9
Share of total demand (%)	16.0	15.9	17.3	18.5				
Renewable								
mtoe	5.0	7.7	14.4	21.8	9.0	6.5	4.3	6.1
Share of total demand (%)	2.2	2.9	4.4	5.5				
Total								
mtoe	229.3	261.5	328.8	400.1	2.7	2.3	2.0	2.3
Share (%)	100.0	100.0	100.0	100.0				

Note: Data for 2010 are estimates. mtoe = million tons of oil equivalent.

Source: KEEI (2006).

million toe to 74.2 million toe during the same period. The consumption of coal, particularly bituminous coal, is predicted to increase from 54.8 million toe in 2005 to 82.1 million toe in 2030.

Projections of CO2 Emissions from Energy Use to 2030

CO_2 emissions from energy consumption are expected to increase from 500 to 790 million tons of CO_2 (tCO_2) between 2005 and 2030, for an average annual increase of 1.8 percent (Table 8.7).

CO_2 emissions have been projected to grow at a relatively high annual average of 2.6 percent between 2005 and 2010, followed by a sharp drop

Table 8.7   Major indicators of energy-related CO_2 emissions in Korea, 2005–30

	2005	2010	2020	2030	Average annual growth rate (%)			
					2005–10	2010–20	2020–30	2005–30
CO_2 emissions (mtCO_2)	500	568	676	790	2.6	1.8	1.6	1.8
Per capita CO_2 emissions (tCO_2/person)	10.3	11.5	13.5	16.0	2.2	1.6	1.7	1.8
Unit change in per capita CO_2 emissions (tCO_2/person)	2.81	3.14	3.68	4.36	2.2	1.6	1.7	1.8
CO_2/GDP (tCO_2/million won)	0.70	0.62	0.51	0.40	−2.1	−2.1	−1.9	−2.1
CO_2/energy (tCO_2/toe)	2.18	2.17	2.06	1.97	−0.1	−0.5	−0.4	−0.4
CO_2–GDP elasticity					0.54	0.44	0.45	0.47

Note: Data for 2010 are estimates. tCO_2 = tons of carbon dioxide. mtCO_2 = million tons of carbon dioxide. toe = tons of oil equivalent. mtoe = million tons of oil equivalent.

Source: KEEI (2006).

to 1.8 percent on average from 2010 to 2020, and further to 1.6 percent on average from 2020 to 2030 (Table 8.8).

For the period 2005–10, the estimated average annual growth rate of CO_2 emissions (2.6 percent) is similar to that of energy consumption (2.7 percent). This relationship is projected to become much weaker during 2010–20 (at respectively 1.8 percent and 2.3 percent) and during 2020–30 (at respectively 1.6 percent and 2.0 percent). This indicates that carbon intensity (CO_2 emissions per unit of energy) is expected to drop significantly after 2010.

In per capita terms, emissions will increase: from 10.3 tCO_2 per person in 2005 to 11.5 tCO_2 in 2010, 13.5 tCO_2 in 2020, and 16.0 tCO_2 in 2030. The carbon intensity of fuel, defined as total CO_2 emissions divided by total primary energy use, is expected to decrease: from 2.17 tCO_2 per toe in 2010 to 2.06 tCO_2 in 2020, and 1.97 tCO_2 in 2030.

Carbon intensity per GDP, defined as the total CO_2 emissions divided by the total national GDP, will decline from 0.70 to 0.40 tCO_2 per million Korean won value-added between 2005 and 2030, decreasing at an average rate of 2.1 percent per year. This trend indicates that less carbon will be emitted in order to realize an equivalent amount of economic value throughout the period. Table 8.8 shows the CO_2 emissions by sector.

Table 8.8 CO_2 emissions by sector in Korea, 2005–30

Energy source	2005	2010	2020	2030	Average annual growth rate (%)			
					2005–10	2010–20	2020–30	2005–30
Transformation								
mtCO$_2$	46.4	57.8	71.4	87.6	4.5	2.1	2.1	2.6
Share of total (%)	34.0	37.3	38.7	40.7				
Industry								
mtCO$_2$	42.7	44.7	50.4	56.2	0.9	1.2	1.1	1.1
Share of total (%)	31.4	28.9	27.3	26.1				
Transportation								
mtCO$_2$	28.7	32.4	40.0	46.3	2.5	2.1	1.5	1.9
Share of total (%)	21.0	20.9	21.7	21.5				
Residential								
mtCO$_2$	13.4	14.3	16.0	17.5	1.4	1.1	0.9	1.1
Share of total (%)	9.8	9.2	8.7	8.1				
Commercial								
mtCO$_2$	5.1	5.6	6.6	7.8	1.8	1.6	1.7	1.7
Share of total (%)	3.8	3.6	3.6	3.6				
Total								
mtCO$_2$	136.3	154.8	184.4	215.4	2.6	1.8	1.6	1.8
Share of total (%)	100.0	100.0	100.0	100.0				

Note: Data for 2010 are estimates. mtCO$_2$ = million tons of carbon dioxide.

Source: KEEI (2006).

INSTITUTIONAL FRAMEWORK

The Korean government established the Inter-Ministerial Committee on the Framework Convention on Climate Change in 1998 under the chairmanship of the Prime Minister. It consists of climate-change-related ministries, national research institutions, and industries. In 1999 the committee adopted the first "Comprehensive National Action Plan for the Framework Convention on Climate Change," which was to be revised every three years. The second and third plans appeared in 2005 and the fourth in 2007. In September 2008, after the introduction of the new administration, the committee adopted the "Comprehensive National Action Plan for Climate Change," which draws on the concept of "low carbon green growth."

The Korean government has long been implementing a wide range of energy conservation policies that have led to considerable GHG emissions reductions. These initiatives have been undertaken because of Korea's extremely high dependence on energy imports, which currently account for 83 percent of energy consumption (excluding nuclear energy). These measures form an integral part of the mitigation strategies for energy security. The plan employs a range of policies and measures aiming at limiting the future emissions of CO_2 and other GHGs from various sources, including the industrial, transportation, residential, and commercial sectors, electricity generation, agriculture, waste management, and the forestry sector.

In February 2009, a new committee called the Presidential Committee on Green Growth was established. This Committee integrates the three former committees: the Inter-Ministerial Committee on Climate Change (chaired by the Prime Minister), the Committee on Sustainable Development (also chaired by the Prime Minister), and the Energy Council (chaired by the President of Korea). The amalgamation of these committees signifies the intertwined nature of the issues that green growth, climate change, sustainable development, and energy present. Note that the committee adopted the "Green Growth National Strategies" and the "Five-Year Plan for Green Growth" in July 2009 (Presidential Committee on Green Growth 2009). The "National Green Growth Strategies" declared three objectives: (1) a low carbon society and energy security, (2) the development of a new growth engine, and (3) enhanced quality of life and an international contribution by Korea, with leadership by Korea. Under these objectives, the government identified ten major agenda items, which can be outlined as follows.

Low carbon society and energy security objective
   1. Build a low carbon society.
   2. Provide greater energy security.
   3. Address climate change.

Development of new growth engine objective
   4. R&D for green technology.
   5. Foster green industry.
   6. Green the industry.
   7. Lay the foundation.

Enhanced quality of life and international contribution by Korea, with leadership by Korea objective
   8. Green transportation and land management.

9. Green lifestyle.
10. Foster international leadership.

The Presidential Committee on Green Growth will serve as the primary institutional body to develop and implement low carbon green growth policies in Korea.

ENERGY POLICIES FOR LOW CARBON GREEN GROWTH

In December 2008 the government adopted the Fourth National Basic Plan for the Rational Use of Energy, 2008–2012 (Ministry of Knowledge Economy 2008a), which is summarized below. Many of these policies are expounded in the National Green Growth Strategies and the Five-Year Plan for Green Growth adopted in July 2009 (Prime Minister's Office 2008 and Presidential Committee on Green Growth 2009).

The Fourth National Basic Plan includes various activities that were implemented even before the plan was adopted. Since it is reformulated every three to five years, however, the plan has also developed new activities specifically aimed at promoting green energy use in various sectors.

The government established an aggressive energy efficiency target. Nationwide energy intensity was to decrease by a total of 11.3 percent from 0.335 toe per million won in 2007 to 0.297 toe per million won in 2012, and by a total of 23.5 percent (down to 0.256 toe per million won) by 2017. This decrease represents average annual rates of 2.4 percent for the five-year period 2007–12 and 2.7 percent for the ten-year period 2007–17, indicating improved efficiency. The government has also formulated a long-term plan for energy efficiency improvements, which aims at decreasing nationwide energy intensity further to 0.185 toe per million won by 2030, representing an overall improvement of 44.8 percent.

Dvelopment of Green Technologies

The government is aggressively promoting the development of energy efficiency technologies that combine information technology and nano technology. By consolidating with information technology, a building energy management system will be developed to reduce building energy use. Efforts are also being made to improve the efficiency of power generation from 38 percent in 2008 to 40 percent by 2012. With nano technology, energy storage technologies will be developed for use in hybrid cars or electric vehicles. R&D funds will be expanded to develop green cars, such as hybrid cars. In addition, highly efficient light emitting diode products will be developed.

The government will support the commercialization of energy-efficient technologies. In particular, emphasis will be placed on seven heavy energy-consuming products responsible for draining large amounts of energy, including boilers, motors, furnaces, drying machines, and lighting appliances. This program is dubbed the Seven Runners Program. In addition, energy efficiency improvements will be pursued for six common home electronics goods and appliances: televisions, refrigerators, air conditioners, washing machines, computers, and set-top boxes. By improving energy efficiency by 20 percent and limiting standby electricity to less than 1 watt, these six consumer goods will become "green" home products.

Energy Efficiency Programs in the Industry Sector

The voluntary agreement program was introduced in 1998 as a partnership program between government and industry, and it has become the main instrument for energy efficiency and CO_2 reduction in the industry sector. The agreement stipulates voluntary targets for energy efficiency or CO_2 reduction through a five-year period for factories consuming 5,000 toe or more per year. Participating firms are provided with low-interest loans, tax credits, and technical support. In the first year, 15 factories participated, and by the end of 2006, a total of 1,353 factories had participated in this program. Among these, the 1,288 factories that participated in 2005 covered 86 percent of energy use in the industrial sector. These factories were planning to invest 5 trillion won over a period of five years, thus conserving 7.9 million toe of energy (or 2.7 trillion won). The voluntary agreement program was expanded to include factories consuming 2,000 toe or more. Among the 1,871 companies that were consuming more than 2,000 toe in 2007, 74 percent (1,383 firms) were participating in the program.

The government is expanding the program to include a negotiated agreement component, with mandatory targets imposed from 2010. Negotiated agreements will be implemented for factories consuming more than 20,000 toe. As of 2007, a total of 385 factories and 29 power plants were consuming more than 20,000 toe. These companies need to employ energy management systems to improve the energy efficiency of their facilities. Box 8.1 shows the performance of voluntary agreement action in 2005; this is expected to save 8 million toe of energy.

Industrial energy audits are conducted to identify the inefficiencies and to provide energy-efficient measures to manufacturers. Activities include the evaluation of the energy efficiency of fuel- and electricity-operated plants, consulting assistance for efficiency enhancement, and

---

BOX 8.1   VOLUNTARY AGREEMENT ACTION IN
KOREA IN 2005

Number of companies   1,288
Energy conservation   7.94 million tons of oil equivalent
Total energy consumption   106.2 million tons of oil equivalent
Investment   5.0 trillion won

Source: Ministry of Trade, Industry, and Energy data, 2007.

---

recommendation of energy efficiency measures. Currently, manufacturers consuming more than 2,000 toe are mandated to undergo an energy audit. The government will increase support for up to 90 percent of the cost of energy audits for small and medium-size factories.

The government has provided low-interest loans for investments by energy service companies, payable in five years with a five-year grace period. In addition, tax deductions are provided both to these companies and to their customers. The coverage of the energy service companies has expanded from investment in lighting systems in the early period to invest-ment in process improvements and waste heat utilization. The government supports these companies by providing them with relevant information on new energy-efficient technologies, and with financial and taxation incen-tives, while inducing investments in energy efficiency and conservation through third-party financing. Investment in the energy service companies amounted to US$106 million in 2006, involving 156 companies.

The government encourages energy supply companies to develop demand side management (DSM) and energy efficiency resource standards programs. Specifically, the government sets mandatory DSM obligations for companies and provides incentives to those successfully fulfilling the DSM obligations. Currently, electricity companies, gas companies, and district heating companies are implementing DSM programs.

Financial assistance to energy efficiency investment amounted to US$512 million in 2007. Low-interest loans are also provided for the installation of energy-saving equipment and district heating facilities, as well as support for energy service companies and the DSM programs.

Energy Efficiency Programs in the Transportation Sector

The government has introduced a fuel-efficiency rating and labeling program that encourages car manufacturers to produce more fuel-efficient vehicles, while providing consumers with better information on the relative fuel efficiency of vehicles. The government plans further expansion of the coverage of the fuel mileage rating system, from gasoline engine passenger cars to other vehicles such as medium and large trucks above 3.5 tons by 2012. In addition, it plans to improve the average fuel efficiency from 12.4 km/liter to 14.5 km/liter by 2012, generating a 16.5 percent efficiency increase for small automobiles (engine capacity below 1,600 cc). The government also seeks to enhance the average fuel efficiency from 9.6 km/liter to 11.2 km/liter by 2012, representing a 16.5 percent increase for medium size automobiles (engine capacity of more than 1,600 cc). Finally, the government plans to use the CO_2 emissions standard as well as fuel mileage standard as a way of increasing awareness of CO_2 emissions from automobiles.

The government adopted measures to promote the use of mini passenger cars with engine capacity of less than 800cc. These measures include tax exemptions and tax deductions, as well as parking and toll fee discounts.

Mass production of hybrid cars began recently in Korea, and the government is taking measures to promote their use. Such measures include excise tax deductions and registration fee exemptions when buying hybrid cars—incentives similar to those currently applicable to small cars with engine capacity of less than 800cc. These measures came into effect beginning July 2009.

The government will take active measures to promote the use of bicycles. Specifically, it will build 9,170 km of bicycle-only roadways and 14,000 bicycle parking lots.

To promote mass transit systems, the government has successfully implemented "bus-only lanes" in the city of Seoul, and will continue to expand bus-only lanes in other major metropolitan areas. In addition, the government is promoting the integration of subway and bus services.

Energy Efficiency Programs in the Commercial and Residential Sector

Buildings in city centers have become larger and more energy-intensive, requiring special energy management. The government promotes voluntary agreements with large energy-intensive buildings (148 buildings in 2007). Technical and financial assistance is provided for the evaluation of energy-saving opportunities and installation of energy saving equipment. The government also promotes the installation of building energy management systems.

The government adopted various programs that encourage households and firms to purchase energy-efficient appliances. The energy efficiency rating and labeling programs will be expanded to include condensing boilers and window fittings. In addition, CO_2 emissions will be marked in parallel with energy efficiency labeling for seventeen home appliances, such as refrigerators, washing machines, and fluorescent lamps. The minimum energy performance standards will be expanded to include three-phase electric motors, adapters, and chargers. The use of incandescent bulbs will be banned by 2013.

Low Carbon Program in the Electricity Sector

The transformation sector, consisting mainly of electric power generation, plays a significant role in reducing CO_2 emissions. In particular, the expansion of nuclear power and natural gas in the power sector is an important option in Korea.

The transformation sector accounted for 36 percent of energy-related CO_2 emissions in 2006. The power sector is expected to remain a major contributor of CO_2 emissions for the next two and a half decades. The nation's total installed capacity of 73 gigawatts (GW) generated 403 terawatt-hours (TWh) of power in 2007.

Nuclear power accounted for the largest share of power generation (39 percent) in 2007, followed by coal (37 percent), LNG (18 percent), and oil (4 percent). Hydropower contributed only 1.4 percent in 2006. From 1986 to 2006, nuclear power was responsible for the largest share in generation. In 2008, however, there was a shift, so that generation from coal power plants accounted for the largest share (39 percent), followed by nuclear power (34 percent), LNG (22 percent), and oil (2 percent).

Power demand is expected to increase from 390 TWh in 2008 to 500 TWh in 2022, representing an average annual increase of 2.1 percent. To meet this demand, the government plans to increase generation capacity from 71.4 GW in 2008 to 100.9 GW by 2022, for an additional generation capacity of 29.5 GW.

As part of its low carbon measures, the government plans to maintain balanced shares of nuclear, coal, and LNG as fuel for the electric power supply. The Fourth Basic Plan for Electricity, 2009–2022 (Ministry of Knowledge 2008b), adopted in December 2008, will increase the share of nuclear power and decrease the share of coal-fired power plants.

In 2009 there were 20 nuclear plants (with 17.72 GW of installed capacity), 46 LNG-fired plants, and 49 coal-fired plants. During the fifteen year period 2009–22, the government plans to build 12 new nuclear plants, 11 new LNG units, and 7 new coal-fired units. In 2008 the shares of nuclear, coal, and LNG in total installed capacity were, respectively, 24.8 percent, 33.2 percent, and 25.2 percent. These shares will change to 27.7 percent,

31.4 percent, and 24.6 percent by 2015, and further to 32.6 percent, 29.2 percent, and 22.9 percent by 2022.

For carbon emissions, power generation is a more important factor than capacity. The share of nuclear power in power generation will increase from 34 percent in 2008 to 48 percent in 2020. Note, however, that the share of LNG will decrease from 22 percent in 2008 to 6 percent in 2020. In addition, the government is promoting DSM programs with a target of reducing peak load by 11.2 percent in 2022.

Low Carbon Program Using New and Renewable Energy

The share of new and renewable energy in total primary energy was 2.4 percent in 2007, including large hydropower plants. This share is very low in comparison with other major countries. In December 2008 the government adopted the Third Basic Plan for New and Renewable Energy Technology Development and Dissemination, 2009–2030 (Ministry of Knowledge 2008c), to promote the use of new and renewable energy. The plan sets targets to raise the share to 4.3 percent in 2015, 6.1 percent in 2020, and 11 percent in 2030 (Table 8.9).

The government will take various measures to promote the use of new and renewable energy. Currently, it is implementing a differentiated feed-in tariff program that guarantees the rates for 15 years for wind and photovoltaic energy. The government will gradually lower the feed-in tariff.

In addition, the government plans to introduce renewable portfolio standards, beginning in 2012, which will require suppliers to generate electricity from new and renewable energies according to certain quotas.

THE WAY FORWARD

Drawing on the most up-to-date studies and government plans, this chapter elaborates on energy policies and programs geared toward a "low carbon green growth" paradigm. Thanks to the establishment of the Presidential Committee on Green Growth, which addresses global climate change and energy risks at the same time, the Korean government is in good shape to promote the concept of "low carbon green growth."

Carbon intensity can be used as an indicator to measure the success of this low carbon economy. In energy terminology, the lowering of carbon intensity requires a shift away from carbon-intensive sources in the energy mix. The expansion of nuclear power and of new and renewable energy is one of the most effective measures toward achieving this goal, and the

Table 8.9 Targets for new and renewable energy in Korea by source, 2008–30

Energy source	1990	2000	2005	2006	Average annual growth rate 1990–2006 (%)
Solar thermal					
thousand toe	33	63	342	1 882	20.2
share of total (%)	0.5	0.5	2.0	5.7	
Photovoltaic					
thousand toe	59	313	552	1 364	15.3
share of total (%)	0.9	2.7	3.2	4.1	
Wind					
thousand toe	106	1 084	2 035	4 155	18.1
share of total (%)	1.7	9.2	11.6	12.6	
Biomass					
thousand toe	518	2 210	4 211	10 357	14.6
share of total (%)	8.1	18.8	24.0	31.4	
Hydropower					
thousand toe	946	1 071	1 165	1 447	1.9
share of total (%)	14.9	9.1	6.6	4.4	
Geothermal					
thousand toe	9	280	544	1 261	25.5
share of total (%)	0.1	2.4	3.1	3.8	
Ocean					
thousand toe	0	393	907	1 540	49.6
share of total (%)	0.0	3.3	5.2	4.7	
Waste					
thousand toe	4 688	6 316	7 764	11 021	4.0
share of total (%)	73.7	53.8	44.3	33.4	
Total					
thousand toe	6 360	11 731	17 520	33 027	7.8
share of total (%)	100.0	100.0	100.0	100.0	
Target in primary energy (%)	2.58	4.33	6.08	11.0	

Source: Ministry of Knowledge Economy (2008c).

government is taking various measures to increase such sources. But there are challenges. Resolving the problem of public acceptance of nuclear power and site issues, for example, is not easy. In the case of new and renewable energy sources, the government needs to deal with the economic disadvantages of investing the necessary large amounts of capital and of creating an initial demand market. When it comes to "green growth," the government faces even more challenging issues. As an export-oriented, small, open economy, Korea needs to come up with ways and means for the nuclear industry and new and renewable energy industry to be competitive in the global market, so that they can export their products. Even though Korea has its own nuclear model, the government has not been successful in selling it in the global market. For new and renewable industries, the country is lagging behind the world's best producers. The Korean government should actively support information technology and nano technology to develop and consolidate new and renewable technologies, as catalysts for the development and deployment of innovative and cost-effective technologies.

There are other indicators for low carbon economy and green growth. Growth requires an increasing GDP, but the economy needs to decouple GDP growth from increases in energy consumption. Energy intensity can be lowered through improvements in energy efficiency, which in turn leads to a decrease in the energy-to-GDP ratio over time. As shown above, Korea has actively formulated energy efficiency policies and measures. There is one caveat for energy efficiency improvement: the government thus far has relied more on regulatory measures (such as voluntary agreements, negotiated agreements, fuel economy standards, and labeling policies), but it could make better use of price mechanisms. For example, in some cases the end-use price is kept low. Currently the price of electricity is lower than the production cost, which can result in over-consumption of electricity. Therefore, the government needs to reconsider its energy pricing policies.

REFERENCES

Korea Energy Economics Institute (KEEI). 2006. *Study on Long-term GHG Projection*. Seoul: Korea Energy Economics Institute.

Korea Energy Economics Institute (KEEI). 2008. *Study on the National Inventory of Korea*. Seoul: Korea Energy Economics Institute.

Ministry of Knowledge Economy. 2008a. *Fourth National Basic Plan for the Rational Use of Energy, 2008–2012*. Seoul: Ministry of Knowledge Economy.

Ministry of Knowledge Economy. 2008b. *Fourth Basic Plan for Electricity, 2009–2022*. Seoul: Ministry of Knowledge Economy.

Ministry of Knowledge Economy. 2008c. *Third Basic Plan for New and Renewable Energy Technology Development and Dissemination, 2009–2030*. Seoul: Ministry of Knowledge Economy.

Pachauri, Rajendra K., and Andy Reisinger, eds. 2007. *Climate Change 2007: Synthesis Report. Contribution of Working Groups I, II and III to the Fourth Assessment Report of the Intergovernmental Panel on Climate Change*. Geneva: Intergovernmental Panel on Climate Change.

Presidential Committee on Green Growth. 2009. *National Strategies and Five-Year Plan for Green Growth.* Seoul: Presidential Committee on Green Growth.

Prime Minister's Office. 2008. *Comprehensive National Action Plan for the Framework Convention on Climate Change.* Seoul: Prime Minister's Office.

9. Issues in establishing a carbon market in Korea

Kyoung-Soo Yoon and Min-Kyu Song

INTRODUCTION

Increasing evidence of global warming and the impact of greenhouse gas (GHG) on it have initiated a worldwide effort to cope with climate change.[1] The 1997 Kyoto protocol was a great step toward multinational collective action to deal with this global phenomenon, regardless of whether its performance has been evaluated as successful. Ahead of the post-Kyoto era beyond 2012, international discussion on global mitigation is now under way and moving toward a certain conclusion. For the scheme to be effective for global mitigation, the post-Kyoto system should be one with more countries involved in it.

Given the current status, it seems inevitable for Korea to play a role in the post-Kyoto system. As is shown in Table 9.1, the volume of carbon dioxide (CO_2) emissions per unit of GDP in Korea is ranked ninth in the world, and the percentage increase from 1995 to 2006 is the topmost among the countries of the Organisation for Economic Co-operation and Development (OECD). Among non-OECD countries, only China and India are similar to Korea in volume and percentage change of GHG emissions. Taking the emissions statistics and the size of the Korean economy into consideration, it is anticipated that some responsibilities for global GHG reduction would be levied on Korea as an advanced developing country in the post-Kyoto system.

The Korean government adopted "green growth" in 2008 as a national development strategy, in preparation for the post-Kyoto era, in need of finding a new growth engine under the changing economic environment, and for the fulfilment of contributions to the international community.[2] The announcement of the Comprehensive Plan on Combating Climate Change (Prime Minister's Office 2008), the Presidential Committee's proposal for the Law for Low Carbon, Green Growth (2009a) and the publication of *Green Growth for Next Five Years* (2009b) as an action plan followed. In addition, the government began planning a midterm

Table 9.1   CO$_2$ emissions: comparisons by region and country, 1990–2006

Region or country	CO$_2$ emissions in 2006 (million tons)	Average annual increase 1990–2006 (%)	GDP (2000 US$ billion using purchasing power parity)	Emissions per GDP (million tons per US$ billion)	Emissions per capita (tons)
World	28 002.7	1.82	57 564.5	0.49	4.28
OECD	12 873.7	0.94	31 157.5	0.41	10.93
Annex I	14 158.1	0.11	31 421.4	0.45	11.18
United States	5 696.8	0.99	11 265.2	0.51	19.00
Japan	1 212.7	0.78	3 538.1	0.34	9.49
Germany	823.5	−0.89	2 254.7	0.37	10.00
France	377.5	0.44	1 695.0	0.22	5.97
Spain	327.6	2.95	1 045.8	0.31	7.44
Turkey	239.7	4.05	576.8	0.42	3.29
Canada	538.8	1.39	1 017.0	0.53	16.52
New Zealand	36.8	3.45	96.7	0.38	8.88
Australia	394.4	2.64	631.9	0.62	19.02
South Africa	342.0	1.86	489.9	0.70	7.22
China	5 648.5	5.94	8 915.7	0.63	4.28
India	1 249.7	4.81	3 671.2	0.34	1.13
Korea	476.1	4.67	1 013.9	0.47	9.86

Note: Annex I countries are the 40 industrialized countries and countries in transition that are signatories of the 1997 Kyoto Protocol.

Source: IEA (2008).

(2020) national target of GHG, based on which the emissions cap and the allocation plan would be set.[3] To do so, three scenarios for the target were proposed with economic impacts, for a consensus-building process. Using the Markal and the computable general equilibrium models developed by the International Energy Agency (IEA) and the OECD, respectively, it is estimated that the volume of GHG emissions in Korea in 2020 will be 37 percent above the 2005 level, under a business as usual scenario, and that a reduction of GHG in a range of 21 to 30 percent would result in a decrease of GDP in a range of 0.29 to 0.49 percent in 2020 (see Table 9.2).[4]

Multifarious mitigation policies are proposed in the series of government plans, including a plan to set up GHG reduction targets by sector, the establishment of an energy target system, policies for renewable and

Table 9.2 *Medium-term reduction targets for CO_2 emissions in Korea:*
three scenarios for 2020

Scenario	Reduction target		Change of GDP from BAU in 2020 (%)
	Change of GHG emissions volume from BAU (813 tCO_2 estimated) in 2020 (%)	Change of GHG emissions volume from 2005 (594 tCO_2) level (%)	
I	−21	+8	−0.29
II	−27	0	−0.37
III	−30	−4	−0.49

Source: Presidential Committee on Green Growth (2009c).

nuclear energy such as the renewable portfolio standard, support for green R&D investment and green information technology, industrial policies to induce the nurturing of green sectors and change of industrial structure, and the vitalization of green finance. Even so, there has been some concern about these policies, in that they include items not yet proven to be related to GHG reduction or environmental improvements, along with concern that those policies are not aligned systematically for the policy objectives.

The concrete mitigation policy, including the establishment of the target or cap of the GHG reduction and the emissions trading system, will be concluded as the current international discussion about the post-Kyoto system is settled. However, it is very necessary at this point to establish the principle under which emissions reduction policies are arranged, and to initiate the design of infrastructure for policy implementation. This chapter examines the desirable direction of the reduction policy package, focusing on the role of the emissions trading system and the design of the trading system to achieve the policy goal.

In the next section, we evaluate the emissions permit trading system as a reduction policy and suggest a way to design it. We also raise a few issues in achieving efficiency and other policy objectives from Korea's perspective, which policymakers should deliberate in designing the policy package. We focus in particular on the design of an optimal policy mix to avoid policy conflicts, on the coverage of participants in the system, on the allocation method for efficient and implementable reduction, and on the way to involve the power generation sector in the emissions trading system. In the third section we discuss the emissions trading system as a financial market, proposing the objectives for the design of the market, characterizing the carbon market in comparison with other financial markets, and make some suggestions for the desirable initial design of a trading system and its infrastructure.

THE EMISSIONS PERMIT SYSTEM AS A REDUCTION POLICY

Alternative Policies and Policy Mix

Climate change due to cumulated GHG is a typical example of negative externality in which current emissions have an effect on the global economy for generations. The reduction of GHG is thus globally efficient when the expected future social cost of adjustment to the climate change is greater than the expected current cost of reduction. The mitigation policy should be designed to minimize the expected cost of reduction. However, the cost of the reduction and, needless to say, the benefit from it are hard to measure and sometimes even difficult to categorize. The reference for this would be a case in which the GHGs emitted are priced and the polluters bear the cost.

When carbon is priced, a firm, as a polluter, will try to minimize the cost in various ways. It will adopt new backstop technologies or better equipment to reduce emissions (the abatement effect), change the composition of inputs (mainly fuels) by taking the carbon price into account (the input-substitution effect), or pass through the price on the products and as a result reduce the outputs (the output-substitution effect). The optimal choice of the firm will be determined at the point where the sum of the marginal abatement cost, the marginal input-substitution cost, and the marginal output-substitution cost is equal to the carbon price.[5]

The best reduction policy will be one providing balanced incentives for these actions. Under information asymmetry between policymakers and economic agents, it is difficult to set direct rules to balance such incentives. Both the carbon tax and the emissions permit trading system are market-based reduction policies in which the marginal reduction cost of each firm in the policy coverage is equalized to the tax rate or the market permit price.

A carbon tax has some advantages in its simplicity, lower transaction cost, and the stability of the price signal, and thus it can be applied more comprehensively. It is also a better option over the permit system with free allocation, in that it avoids the political difficulties in the allocating process,[6] and raises revenue to create the "recycling effect" (Bovenberg and Mooij 1994; Kim 2007). Indeed, most general equilibrium models in recent research calibrate that a carbon tax minimizes the decrease in production relative to other policies due to the revenue recycling effect (see for example Cho 2008 and Goulder et al. 1999). On the other hand, however, a permit trading system has some advantages relative to a carbon tax. First, if large asymmetric information exists in the reduction cost of firms and if the heterogeneity of firms is not negligible, the permit trading system would be better, in that the price is discovered through the market mechanism, and not just set by policymakers (Newell and Stavins 2003; Stavins 1995). Second, the authority can design the decreasing trajectory of the emissions cap over a period of years under the emissions trading system, whereas it is not easy to change the carbon tax rate flexibly. Third, the permit trading system is much easier for the linkage to other countries' reduction policy, where national targets and thus the carbon prices differ. International carbon trade with terms-of-trade set appropriately, or through an offset program, enhances global efficiency (Kim and Jang 2008).

In comparing the carbon tax and the permit system, one important issue is which scheme is better under uncertainty. GHG remains in the atmosphere more than a hundred years, so the benefit from reduction is marginally small in the short time horizon. By contrast, the uncertainty of marginal abatement costs in the short time horizon is huge: that is, the marginal cost is relatively steep compared with the marginal benefit in the policy-relevant time horizon. In this case, quantity regulation (the permit system) causes potentially more burden than price regulation (a tax) (Weitzman 1974; Stern 2007; Hahn and Kim 2008). However, the permit trading system is a very flexible scheme that could be modified to resolve the problem. Versions of the permit system have been proposed for this purpose, including a scheme with a combination of long-term and short-term permits (McKibbin and Wilcoxen 2002), a scheme with a "safety valve" in which a price ceiling is set (Pizer 2002), and a scheme with "allowance reserves" in which a price ceiling is applied partially but a long-term reduction target is committed (Murray et al. 2009). Banking and borrowing of permits would also help to reduce the intertemporal uncertainties. Those schemes are intended to make the vertical permit supply in the pure quantity restriction to be flexible. More generally, the authority (say, a carbon central bank) could play a role in issuing (allocating) and secondary (trading) markets to reduce the cost from uncertainty; committing to mid- or long-term reduction targets, it could control the permit volumes in each period, taking temporal shocks and updated information into account.

Standards-based policies are implemented by command-and-control, such as efficiency standards for appliances, vehicle fuel-economy standards, and renewable portfolio standards for electricity generators. They have some shortcomings, when compared with market-based policies, since they usually provide incentives for a specific action and may distort incentives for the optimal choice by economic agents. Various subsidy policies share this aspect. This problem becomes amplified when the asymmetric information between decisionmakers and policymakers is huge. It would be desirable for such policy instruments to be used only when the market-based policy does not work.

Another set of policy instruments is related to technology and innovation. It is well known that there are market failures in this area due to, among other things, knowledge externality, adoption externality, and incomplete information (Jaffe et al. 2005). Without these externalities, investments in technology or innovative adoptions will be rewarded both from the goods market and from the carbon market. The technologies related to the GHG reduction, however, usually exhibit a wide range of uncertainty both in technical success and marketability, which implies that market failures may be strong. R&D support and other subsidy policies to enhance the innovation could be selected to cure these market failures. Though policymakers have a tendency to prefer these policies, owing to their temporal effect on macroeconomics and political easiness, they possibly result in over-investments, or reward firms doubly, transferring wealth from taxpayers to shareholders of firms. It is important, therefore, to design the policy so that limited resources are used where market failures are serious and so that the economy enjoys the spill-over effect from the successful output.

Owing to the multidimensional market failures discussed above, the optimal design of reduction policy might include more than one of the policy tools. In the case a policy mix is needed, the analysis of policy interaction is crucial. The scope, objective, and the way of operation and implementation should be considered for the analysis of direct, indirect, and trading interactions between policy instruments (Sorrell and Sijm 2003). This analysis becomes more difficult as the number of instruments increases. One way to make it easy is to set a reference policy instrument with which other instruments are compared, to avoid the double regulation and double crediting problems. A reference policy will be better if it is flexible, stable, easy to combine offset programs, and appropriate for international linkage. Market-based policies usually have these properties.

If the permit system is selected as the basis of the policy package, considering the international linkage, it is probable that only some of the firms will be able to participate in the system, owing to the transaction cost. In this case it is desirable that other emitters in the economy be covered by the carbon tax in a fair and efficient policy package. The tax rate could be determined in reference to the permit price, and firms on the border between the two policies may be allowed to select one of them. To lessen the disadvantages of the permit system, both flexibility and consistency should be taken into consideration in designing the initial allocation process and the trading market.

In the following subsections, we raise a few issues about the design of the permit trading system.

Participants and Allocation

One of the issues in the introduction of the permit trading system is to decide the coverage of participants among economic agents, which is related to the point of regulation. This determines the range of GHG emissions covered by the system of total emissions.[7] Two options are suggested: the upstream sector approach and the downstream sector approach.

In the upstream approach, participants will be firms producing, importing, or selling fossil fuels, and the emissions cost is shifted to downstream firms through the price mechanism. The advantage of this approach lies in its comprehensiveness. It is desirable for the reduction objective since it covers almost all GHG emissions, and most emitters are affected by the system. It also minimizes the administrative cost because of the small number of participants, but this also implies (as a mirror image) that the trading market may not work efficiently and fairly. This is particularly the case in Korea, where upstream energy markets are characterized as monopolistic or oligopolistic, and most of the fuel supply depends on imports. Natural gas is supplied almost entirely by a monopolistic firm under strong price regulation, and the electricity market is similar, as we will discuss in the next subsection. In this case, appropriate shifts of permit costs to the energy retail prices appear hard to achieve. The wholesale oil market is quite oligopolistic, and potentially collusion could matter.

The downstream approach is based on the polluter-pays principle in which economic agents that emit GHG by combusting fossil fuels pay the emissions cost. Though it induces reduction actions directly, and many participants in the system make the trading market work more efficiently, transaction costs and administrative costs will prevent the system from encompassing all emitters. Instead, sizable agents emitting GHG above a certain level—for example above 2,000 tons of oil equivalent in the beginning stage—could participate in the system. In this case, other policy instruments such as a carbon tax, a standards-based policy, or offset programs for small emitters should be combined to avoid carbon leakage.[8]

International trends and the post-Kyoto protocol will be critical in selecting the point of regulation for Korea. If the international emissions trading system is envisioned and the protocol is similar to that of the

European Union's Emissions Trading Scheme (EU ETS), the downstream approach will be the option to select.

Another important issue in designing the system is the method of allocating permits.[9] In a standard theoretical model, the way permits are initially allocated does not affect the prices of final goods to which the emissions cost is passed on (see for example Montgomery 1972). Each firm in the emissions trading system perceives the market permit price as an opportunity cost, whether it actually purchases the permit in the market or the permit is granted freely. This implies that as long as the total amount of permits distributed is socially optimal, efficiency is achieved regardless of the initial allocation method. Distribution, however, is affected: free allocation violates the polluter-pays principle, generating a lump-sum transfer of wealth from tax-payers to shareholders of firms.[10]

There are some conditions under which the equivalence holds. If only a small proportion of firms participate in the system, if decisions about entry and exit are affected by financial constraints, if there are considerable market powers either in the product or the permit trading markets, or if product prices are under regulation, free allocation distorts firms' decisions and thus affects market efficiency. Free allocation also removes the opportunity for revenue recycling, making the system less efficient than a carbon tax or an emissions trading system with an auction (Cramton and Kerr 2002).

Another problem with the grandfathering method is in terms of dynamic efficiency. In grandfathering, permits are usually allocated on the basis of the firm's past emissions history, resulting in heavy emitters being allocated more. On the firm's side, this lessens the burden and the shock from the introduction of the system, achieving cost efficiency given a fixed industry structure and technology level. It does not, however, count the broad concept of emissions costs, and thus does not induce change in industry structure or the creation of new markets. Combining other policy instruments such as subsidies for R&D investment and purchase of environment-friendly products with a free allocation emissions trading system may help, but there is a concern about government failure.

Taking the political issues into consideration and to reduce the shock from the change of industry structure, policymakers may design a hybrid model. In the model, auction and free allocation are used simultaneously in the beginning stage, but in which the trajectory of the composition of allocation methods, increasing the portion of the auction, is proposed. The portion of free allocation in the early period can be viewed as the compensation for the decline in corporate value from the introduction of the regulation. In this case, the time consistency and commitment power of the government policy would be very crucial.

Power Market and Carbon Market

In the case that an emissions trading system is introduced in Korea, power generation companies are expected to be major players. As shown in Figure 9.1, the power sector is estimated to account for around 30 percent of total GHG emitted in Korea, and thus around half of the permit would be issued to and exercised by the power sector, at least at the beginning stage of the emissions trading system. In addition, in the sense that the power sector supplies the economy a source of energy, as well as consuming fossil fuels, special consideration should be given in designing an emissions reduction policy package. The current industry structure of the power sector, however, leaves a lot of concern about whether a market-based reduction policy will work efficiently. In this subsection, we overview the industry structure of the Korean power sector, investigate factors conflicting with the efficient operation of an emissions trading system, and then suggest options to avoid this problem.

Traditionally, the Korean power market was under strict regulation in which the vertically integrated Korea Electric Power Corporation (Kepco), overseen by the government, played a monopolist role in all power generation, system operation, and sales. From the late 1990s, the Korean government carried out a liberalization of the power market. As a result, in the upstream market, six electric power companies were separated from Kepco as subsidiary companies. Among them, one is for the nuclear and hydroelectric power plants, and the others are mainly for thermal power generation. The government planned to sell the companies to the private sector for the market to be competitive, but up to now such plans have not been successful. Several private companies are also supplying electricity, but the proportion is still small.

Currently, Kepco is a system operator and a monopolistic marketer and distributor in the power market. In the upstream wholesale market, Kepco acts as a single buyer, purchasing electricity through the Korea Power Exchange (KPX). The operation of the purchases uses the mechanism of a "cost based pool" (CBP), in which each power generating company reports the available capacity and KPX determines which facilities come into operation, based on the demand forecast and information of the cost of each facility. For the stability of the electricity price and to prevent excessive profits for generating companies, the market is divided into two parts: a base-load market for coal, nuclear, and hydro-power generators and a peak-load market for generators using liquefied natural gas or oil. The highest cost of the power generator is called the "system marginal cost" (SMP), which determines the (peak-load) market price, while for the base-load generators, the "base-load marginal price"

million tons of CO$_2$

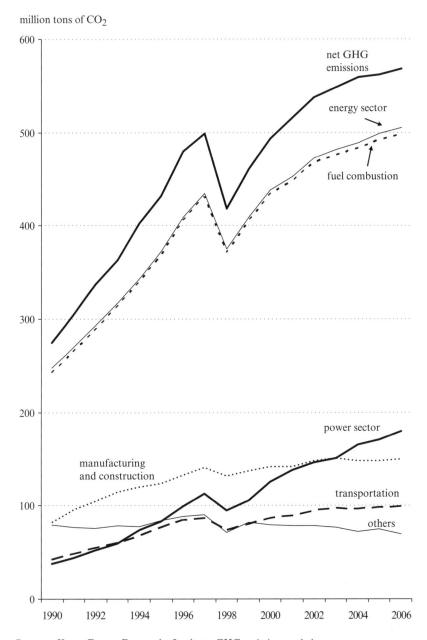

Source: Korea Energy Economics Institute, GHG emission statistics.

Figure 9.1   GHG emissions in Korea by sector, 1990–2006

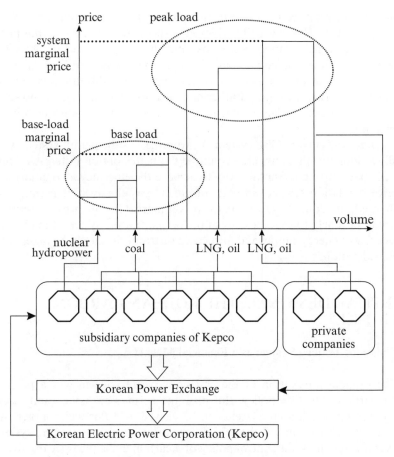

Figure 9.2 Structure of Korea's upstream power market

(BLMP) is determined separately. This structure is depicted in Figure 9.2.

This process differs from the "price based pool" (PBP), in which both price and quantity are in the bidding system. The CBP scheme was adopted in the Korean upstream power market because of the concern about collusion in auctions. In the PBP system, if it is not sufficiently competitive, generation companies would want to collude by submitting high bids, turn by turn, to elevate the SMP and thereby increase the collective profit.[11]

The compatibility of the CBP system and the carbon market system as a reduction policy is doubtful, especially when permits are allocated by grandfathering. For the permit system to work effectively, it is critical that the permit price should be passed through to the electricity price in an adequate way. This is usually achieved in such a way that the permit price is considered as an opportunity cost, but this will not work in the CBP market.[12] Without adequate pass-through, there would be distortions in decisions about the composition of fuel inputs. In addition, there might be entry barriers with grandfathering allocation, which might contradict the government's policy to encourage competition by inducing private entrants.

Under the current CBP system, a carbon tax or the permit system with allocations through auctions would be the better option in terms of efficient pass-through. Another way to achieve the efficiency is to adjust the price regulation scheme, so that consumers pay the average permit cost charged. In addition, reforms in the power market to enhance competition would be a help to the efficiency of the permit trading system. Finally, the security of energy supply and fairness in distribution should be taken into consideration.

INITIAL DESIGN OF THE KOREAN CARBON MARKET

Emissions Trading System as a Financial Market

The original motivation of the emissions permit system was a part of the carbon reduction policy. But once the permits are issued and ready for trading, the trading system can be viewed as a financial market. As described above, one of the advantages of an emissions trading system over a command-and-control policy or a carbon tax is the price discovery through the market mechanism, from which underlying benefits-costs of mitigation are revealed in an efficient way. This price signal plays an important role in decisionmaking by various economic agents, including firms, consumers, investors, and policymakers. By contrast, if the market price is overly volatile, the system may cause additional risks for the mitigating firms.

A well-functioning price mechanism is, therefore, a key factor for the success of the market. We suggest that a target in the design of the carbon market should be to advance the roles of price discovery and risk management; and, of course, each of these two assists the other. This section tries to identify factors that can promote these two essential functions in carbon markets. We first focus on the intrinsic structures and characteristics of carbon markets under the cap-and-trade system, and associated derivatives markets are mentioned.[13] A few recommendations to policymakers

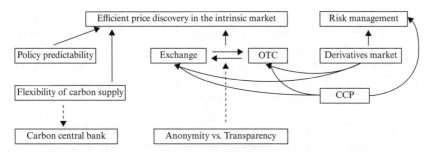

Figure 9.3   Policy objectives and considerations in carbon-market price discovery and risk management

follow, with these factors in mind. The objectives, factors, and their relationship taken into account in this section are summarized in Figure 9.3.

Characteristics of Carbon Markets

Given the carbon reduction target and allowances set by the authorities, incentives for demand and supply in a carbon allowance market are very simple. The buy-sell-keep decision of each firm in the system depends on the marginal abatement cost and the market price, and the market clears at the point at which the demand and supply coincide. At market equilibrium, quantities of demand and supply are equalized, and the marginal cost of a critical firm on the verge of demand and supply comes to be a market price of allowance. Changes of the underlying environment, such as the abatement cost or the cap, affect the market equilibrium. For example, it is easy to see that the overall rise of the reduction cost among the participating firms increases the market price of allowance, while it does not affect quantities demanded and supplied. If the reduction target increases, the market price will rise, and demanders in the market will become fewer due to the permit constraint allotted; the demand is then concentrated into a smaller number of firms. And if the penalty is not high enough for firms that do not satisfy the reduction target, the intrinsic transactions will be crowded out (along with the subsequent transactions), resulting in the disappearance of the carbon market.

Like other securities markets, the carbon market is affected by firm profitability and various factors in the financial markets. For example, if the market interest rates rise, if stock markets deteriorate, or if firms become less profitable, then the costs of carbon reduction will increase, and the price of the allowance will be higher.

A few notable things, however, distinguish the carbon market from other well-known securities markets, such as a stock market or a bond market. First, the intrinsic demanders and suppliers are limited to the firms under the cap-and-trade scheme. This is because other agents such as individuals and financial intermediaries have no reason to buy and hold the allowances unless they are forced to meet the cap. Of course they may trade for temporary reasons, but even temporary trading would not take place without the intrinsic transactions. Second, there is a tendency for the demand to become concentrated into a smaller number of participants as the reduction target becomes higher—which will be the case when the carbon authorities keep lowering the emissions ceiling and raising the reduction level over time. The reason for this is as follows. Since demand and supply of allowances are determined in comparison with the reduction costs, an increase in the reduction requirement places a heavier burden on the firms that are less cost efficient. Third, a penalty is usually imposed for excess carbon emissions. It will (or should be) high enough for the existence of the market: otherwise, the incentives to trade the allowances in the market will disappear.

These distinctive characteristics of carbon markets are important considerations when assessing the macro-structure of a carbon market and designing a newly launching carbon transaction system. More concretely, the fact that fundamental suppliers and demanders are limited to participants of the cap-and-trade system, along with the tendency for demand to become concentrated among a smaller number of firms, implies that "anonymity" of the carbon transactions is needed, more or less.

Transaction anonymity is a trade-off of market transparency from a general perspective. Many theories indicate that transparency improves the fairness of market transactions, because it makes information about transactions and underlying assets more accessible. (Kim et al. 2009 summarize this issue comprehensively.) But, this is not always the case, especially when it applies to informed dealers and large institutions. They may not want too much information about themselves to be revealed in the market. If they do not, transparency of the market can sometimes be a factor denting the liquidity of the market (see for example Bloomfield and O'Hara 1999). This is a possible scenario because dealers and large institutions are frequently major liquidity providers in markets; if their transaction incentives are damaged, the whole liquidity of the market can be threatened.[14] Anonymity of the market is explained by reversing the market transparency. In other words, market anonymity can improve the trading incentives of the so-called informed traders at the cost of uninformed traders' incentives. If not enough anonymity is maintained in a market, the strategic behavior of the uninformed, which is based on the behavior of the informed traders, will cost the informed the benefit of their

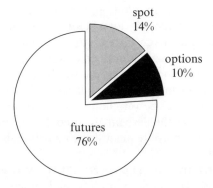

Source: ECX (2008).

Figure 9.4 *Market shares by instruments in the European Union's Emissions Trading Scheme, September 2008*

tradings. Thus there could be an optimal level of anonymity and transparency in a market.

Because the intrinsic demanders and suppliers in a carbon market are the informed traders with their own information and are relatively fewer in number, if there is not enough anonymity, the cost can be detrimental to the market. A contraction in the number of demanders over time. with the lowering of the emissions cap, is an additional reason for anonymity. In any case, there should be some room for market anonymity if we want vivid transactions by the intrinsic participants. It is therefore a good idea for the carbon authority to admit the varying transparency and anonymity in the market, for example by not forcing every carbon transaction only in an exchange market. The over-the-counter market should not be avoided.

For further characteristics of the carbon market, we can refer to the experience of the EU ETS, the largest carbon market in the world. One notable feature of carbon markets is that the derivatives are more active than the underlying assets—for example, the European Union Allowance (EUA) or certified emissions reductions (CERs) themselves in the EU ETS. The trading volume of carbon derivatives is said to be more than 85 percent of the total carbon trading, as shown in Figure 9.4. The BlueNext seems to be the only exchange that is dealing with spot transactions of the EUA and the CERs. The European Climate Exchange (ECX) has introduced one-day carbon futures rather than spot trading. The spot trading underlying carbon products corresponds to intrinsic transactions. It can be guessed easily that the spot trading volume would be limited, because the intrinsic trading incentives come mainly from the firms under the cap-and-trade scheme. In addition, even the firms under the scheme have little incentive to trade very long before the enforced reporting time of emissions obligations. Even if they want to trade very early from the reporting time, they will use futures or forward contracts, rather than the EUA or the CERs themselves, because the derivatives markets are more liquid and entail lower transaction costs. For the above reasons and others, it is thought to be very natural that the underlying assets are traded less actively than the derivatives. It could be conjectured that the spot trading would be more frequent if the reporting time differed across the firms.

By contrast, derivatives trading is very active for reasons of risk management associated with carbon trading and others. As an instance, we can think of derivatives demand to handle the various uncertainties during the processes of carbon reduction investments. The derivatives are very efficient tools for coping with uncertainties about carbon prices and abatement costs, because of lower trading costs and higher leveraged effects than the spot trading. Besides the need for derivatives by the intrinsic participants, many others want to trade in the derivatives markets, such as dealers and intermediary institutions that have to manage their carbon-related positions and carbon funds. And there will be arbitrageurs and speculators in the derivatives markets. Summing up, the derivatives markets are widely and actively utilized for various purposes, because they are efficient risk management tools and very liquid markets. This fact makes the derivatives markets more attractive.

Second, the trading portion in the over-the-counter (OTC) markets is very high: 58.1 percent of the EUA trading volume is in the OTC markets and 87 percent of the CERs (see Table 9.3 and the orbeo website). Even though the carbon products are well defined and standardized, the OTC markets play non-negligible roles. A few reasons can be suggested. As in the previous subsection, the anonymity of transactions is a strong candidate. Inherently, the OTC markets are able to supply various levels of anonymity and to do so differently from the exchanges. And the OTC carbon markets do not require very high search costs, because the intrinsic participants in carbon markets are relatively few in number. Neither do they have much need of continuous trading, because the intrinsic transactions tend to be rushed and clustered around the reporting periods. Especially, the widely used centralized counter parties (CCPs) deserve topmost credit. The Intercontinental Exchange (ICE) and the London-based clearing house LCH.Clearnet, for representative examples, provide CCP services even to OTC transactions. This lessens the counter party risks involved with OTC and makes the CCP services from the exchanges less attractive. Even for futures contracts in the BlueNext, the delivery timing is the

Table 9.3   *Shares of carbon market transactions: exchanges versus over the counter markets and centralized counter parties versus non-CCP services (%)*

	Exchange	Over the counter	Total
European Union allowance	41.9	58.1	100.0
Certified emissions reductions	13.0	87.0	100.0
	CCP	Non-CCP	Total
Over the counter	73	27	100
Exchange	100	0	100

Note: Primary CER.EUR and other voluntary markets are excluded.

Source: Author's calculations based on orbeo website.

same with the OTC transaction linked to LCH.Clearnet (see the BlueNext website).

Third, the CCPs are utilized very well. The exchanges are supplying the CCP services by themselves or through association with external CCPs. The ECX and the BlueNext are associated with the ICE and LCH. Clearnet, respectively. The ICE and LCH.Clearnet provide CCP services for OTC trading as well. Ninety-two percent of the EUA OTC trading and 75 percent of the CER OTC trading are passing through the CCPs.

The CCPs service the economic benefits regarding the efficiency of transaction record-keeping and the risk management of counter-party risks. This is a critical component in carbon markets to maintain and manage the transaction logs, describing "who holds the carbon and in which country?" The relation with CCPs and registries helps to prevent errors and omissions in records. CCPs, local registries, and the community independent transaction log are tightly connected, in the case of the EU ETS, for keeping the transaction logs. The OTC markets may enjoy their dependence on the CCPs, because they do not then have to go to the registries separately. More fundamentally, the CCPs eliminate the counter-party risks in transactions, which enhances the marketability of carbon products. The CCPs are supporting both exchange and OTC trading through efficient record keeping and the elimination of counter-party risks.

Suggestions for Newly Launching Carbon Markets

For the emergence of the Korean carbon market, the characteristics we have described above suggest some issues to be considered. First, the carbon authority should enhance the predictability of reduction targets over time. The reduction target set by the authority is a key factor in the equilibrium conditions of the carbon market. If the carbon reduction investment takes a fairly long time, the unexpected reduction target will generate unnecessary price distortions and a socially inefficient level of reduction efforts. Concretely, if the reduction target is underestimated, overall reduction investment falls short, and the imbalance of demand and supply will force up the price. If it is overestimated, the abatement investment is above the desirable level, and the price falls excessively. Therefore, in order to attain socially efficient carbon abatement and price formation, it is essential to make the reduction targets very predictable.

Second, it is a good idea to encourage the derivatives markets for the fast establishment of the overall carbon market. The original idea of a carbon market was to employ the price mechanism for efficient emissions abatement. That is, the essential issue in a carbon market is how to make the market find the price efficiently. In general, liquidity in the market and continuous formation of prices are desirable for efficient price-discovery. But the intrinsic trading volume is limited by the total caps, and the incentives for frequent intrinsic trading are deficient—which are inherent barriers to efficient price-discovery. By contrast, the derivatives markets are more liquid and more continuous potentially, because of lower transaction costs, higher leveraged effects, and various types of participants. It is more probable that the derivatives markets will strengthen efficient price-discovery. Additionally, active derivatives markets would contribute to the development of the underlying carbon markets. These would be chances for risk management related to the cap-and-trade scheme and to minimize the effects of unexpected shocks. Derivatives markets of high quality would help to handle the unpredicted changes of the caps. Dealers, intermediary institutions, and carbon funds could manage their positions using the derivatives markets. We suggest that the underlying carbon products and their derivatives should be advanced together, for the synergy effects, rather than the approach of underlying products first and derivatives after.

Third, it is recommended that carbon trading should not be forced only in the exchange and that OTC functions be respected at the same time. We would rather give the market participants free choices to select either the exchange or OTC transactions. Despite the standardization of the carbon products, market participants want various degrees of anonymity and transparency. It is preferred that the market freely choose its type of trading for each case. The weaknesses of the OTC markets, such as record keeping and counter-party risks, can be supplemented by the extended usage of the CCPs.

Fourth, the CCP is an excellent addition to the market structure from the beginning stage. The CCP would strengthen market stability by eliminating the counter-party risks and improve the efficiency of keeping the transaction logs. And such functions of the CCP would be prominent in the OTC transactions. It is recommended that the CCP employ an international protocol that can be extended easily to the international transaction log, because the domestic registry should be linked to the international one at length. The capital adequacy requirement of the CCP should be considered, because the counter-party risks are condensed into the CCP.

Fifth, a carbon central bank is well worth pursuing. One of the problems in carbon markets is the instability of price. For example, carbon prices in the EU ETS are highly volatile around April and May, and suddenly fall during its first phase, which makes people doubt that the market is finding the price efficiently. The inflexible supply of the carbon products in quantity and timing is often pointed out as the reason. This implies that a third party might be needed for the improvement of the market. The party could buy the carbon if the market environment deteriorates seriously, and sell if the market is overheated. It can also operate the issuing market through auction, as discussed in the previous section. This function would be a counterpart to the central bank with respect to the money supply.[15]

CONCLUSION

In this chapter we outline the prospects for a Korean carbon market system as a part of an emissions reduction policy package and as a financial market, making some policy recommendations. Although a comprehensive market-based reduction policy through an emissions permit trading system would be the most efficient course, a policy mix may be inevitable under conditions such as market failure in innovation and technology diffusion, price regulation on energy, severe transaction costs, and so forth. The policy package should be designed to balance rewards for reduction actions. It is recommended that the permit system should play a central role in the policy package, working as a reference based on which other policies are evaluated and executed.

Special attention should be paid to the power market in that power plant companies will be major participants in the emissions trading system. However, a few factors will hinder the efficient working of the market. The CBP system in the Korean upstream power market and widespread price regulation not only reduce the incentive to reduce emissions, but also hamper the pass-through of the permit cost onward to the electricity price. Reforms may be needed in the industry structure, if the price mechanism is to work effectively in both the electricity market and the carbon market.

The permit trading market should be designed to help the reduction objectives through price discovery and the provision of risk management tools. A balance among transparency and anonymity, spot and derivatives trading, and the exchange and OTC markets should be taken into account. Predictable reduction targets over time, inducing active derivatives markets, parallel developments of the exchange and OTC markets, and possible additions of CCPs and a carbon central bank to the system are recommended.

ACKNOWLEDGMENTS

We thank Chin Hee Hahn, Stephen Howes, Jaehoon Kim, Dongsoon Lim, Suil Lee, James Roumasset, Jos Sijm, Taehoon Youn, and other participants in the EWC-KDI conference, for their comments and advice.

NOTES

1. For a detailed discussion of scientific findings of global climate change, see IPCC (2007). In Korea, it is estimated that the temperature rose by 1.5° C between 1906 and 2005 (compared with a global average of around 0.74° ± 0.18° C) and that the annual average precipitation also increased, from 1,166 mm in 1920 to 1,501 mm in 2006. Unlike other countries at a similar latitude, the climate change of Korea is not toward desertification but toward subtropical conditions.
2. The economic concept of "green growth" integrates economic growth, sustainability, and coping with climate change. Hahn and Kim (2008) define it as "growth reducing the gap of per capita GDP with advanced countries, participating efficiently and fairly in the international effort on climate change."
3. There has been concern that it is not a good strategy for the government to announce the target before an international agreement for the post-Kyoto period is settled. However, setting the target seems consistent with the negotiation strategy of the Korean government, in which developing countries commit to nationally appropriate mitigation actions (NAMAs) and report their mitigation actions to the NAMA registry.
4. The assumptions used to estimate business as usual GHG emissions and the impact of GHG reduction on GDP are not fully open to the public. A few potential problems should be noted, among others that might affect the estimation result. First, the GDP growth rate used in the estimation does not fully reflect the economic downturn after 2007, and thus the business as usual emissions volume in 2020 might be overestimated. Second, future figures of the share of the manufacturing sector in GDP could have a big effect on the result, but the share is very difficult to predict, since it is higher in Korea than in any other OECD country at the current time. Third, information about current and future reduction technologies would be gathered from industry, and thus their cost could be over-reported.
5. At a macroeconomic level, the direct effect of this carbon constraint will reduce the aggregate output. On the other hand, it will create a new market for intermediate goods with new technology and new consumption goods—a market that could not be opened without the constraint, which in turn might help the growth of the economy, referring

to the endogenous growth theory. In addition, when the pricing is directed to the government revenue, the constraint generates a "tax interaction effect" and a "revenue recycling effect," which affect the growth of the economy. The overall macroeconomic effect of this constraint will be determined by how these indirect effects offset the direct effect.

6. While the political process for the allocation of permits incurs social costs, the politics could be a merit for the permit system, especially to design a second-best policy when the first-best is not politically viable. See Stavins (2009), where the author evaluates the Waxman-Markey bill in terms of how it avoids a "political-giveaway" in the proposed permit system in the United States.

7. Another issue on the coverage of GHG emissions is what kind of GHG is to be included in the system. There is a consensus on this issue in Korea that, in the beginning stage, only CO_2 would be included, due to the ease of measure, report, and verification, but that other GHGs should be included as the measure, report, and verification system is developed further. The carbon offset program can be used before other GHGs are included.

8. The electric power sector has a special position, because firms in the sector consume fossil fuels but at the same time supply energy to other firms and households. We examine issues related to the power market in a later subsection.

9. The discussion on the allocation mechanism here is for the national allocation plan (NAP) and is based on the assumption that a national cap is imposed in the post-Kyoto system, or that the government will set a cap voluntarily. If, for example, the sectoral approach is adopted in the post-Kyoto system, the allocation method should be investigated from different perspectives.

10. Regulation of pollution emissions itself affects the distribution regressively. See Dinan and Rogers (2002). Free allocation may aggravate this effect.

11. This strategic behavior may occur even in the CBP market in the form of the moral hazard. It is argued that, in a CBP market, the incentive to lower the fuel cost is weak. This is especially the case when each company operates both base-load generators and peak-load generators. See Do and Kim (2004).

12. The "windfall profit" of generation companies, a problem frequently raised in the EU ETS, will not happen in the CBP system. In fact, there is a tradeoff between the windfall profit problem and the efficient pass-through of the carbon cost, since the windfall profit occurs due to the gap between the opportunity and the accounting costs.

13. By "intrinsic" trading we mean transactions among the participants who are initially allotted carbon allowances and obliged to satisfy their carbon caps or ceilings set by the carbon authorities. The intrinsic transactions are contrasted with the other subsequent transactions, which would not result if the intrinsic trading does not take place.

14. For this reason, the Securities and Investment Board in the United Kingdom points out the necessity of limiting the market transparency more or less for its liquidity.

15. To achieve stability in the carbon price, a few other options could be thought of. Use of perpetuity, allowing banking, and active derivatives markets would help to make the price less volatile through inter-temporal transactions. Diversifying the time of issue or reporting through a year can also be an option to reduce seasonal amplification of volatility. These options, however, should be taken after careful consideration of side-effects.

REFERENCES

Bloomfield, Robert, and Maureen O'Hara. 1999. Market Transparency: Who Wins and Who Loses? *Review of Financial Studies* 12 (1): 5–35.

BlueNext. Website www.bluenext.eu.

Bovenberg, Ary Lans, and Ruud Aloysius Mooij. 1994. Environmental Levies and Distortionary Taxation. *American Economic Review* 84 (4): 1085–89.

Cho, Gyeong Lyeob. 2008. Economic Impact of Green-house-gas Reduction. Paper presented at the Symposium on Green Growth: Groping for the National Growth Strategy, held by the Korea Development Institute, the Presidential Council for Future and Vision, and the National Research Council for Economics, Humanities, and Social Sciences. In Korean.

Cramton, Peter, and Suzi Kerr. 2002. Tradeable Carbon Permit Auctions How and Why to Auction Not Grandfather. *Energy Policy* 30 (4): 333–45.

Dinan, Terry, and Diane Lim Rogers. 2002. Distributional Effects of Carbon Allowance Trading: How Government Decisions Determine Winners and Losers. *National Tax Journal* 55 (2): 199–221.

Do, Hynjae, and Kyjoon Kim. 2004. A Study on the Incentive to Curtail the Generation Fuel Cost in CBP Market Competition. *Energy Economics Research* 3 (2). In Korean.

European Climate Exchange (ECX). Website www.ecx.eu.

Goulder, L., W.H. Parry, R. Williams III, and D. Burtraw. 1999. The Cost Effectiveness of Alternative Instruments for Environmental Protection in the Second-Best Setting. *Journal of Public Economics* 72 (3): 329–60.

Hahn, Chin Hee, and Jahoon Kim. 2008. Green Growth National Growth Strategy: Concept, Framework, and Agenda. Paper presented at the Symposium on Green Growth: Groping for the National Growth Strategy, held by the Korea Development Institute, the Presidential Council for Future and Vision, and the National Research Council for Economics, Humanities, and Social Sciences. In Korean.

Intercontinental Exchange (ICE). Website www.theice.com.

Intergovernmental Panel on Climate Change (IPCC). 2007. *Climate Change 2007: The Physical Science Basis. Contribution of Working Group I to the Fourth Assessment Report of the IPCC.* Geneva: Intergovernmental Panel on Climate Change.

International Energy Agency (IEA). 2008. *IEA Statistics.* Paris: International Energy Agency.

Jaffe, Adam B., Richard G. Newell, and Robert N. Stavins. 2005. A Tale of Two Market Failures: Technology and Environmental Policy. *Ecological Economics* 54 (2–3): 164–74.

Kim, Seung Rae. 2007. *A Study on Tax and Fiscal Policy for the Introduction of Carbon Tax.* Seoul: Korea Institute of Public Finance. In Korean.

Kim, Yong-Gun, and Kibok Jang. 2008. *Economic Impacts of International Greenhouse Gas Emissions Trading.* Research Paper 2008 RE-11. Seoul: Korea Environment Institute. In Korean.

Kim, Youngmin, Jonghyun Park, and Moonsung Hwang. 2009. Impact on the Korean Economy of Responding to Climate Change and Implications. *Bank of Korea Monthly Bulletin* (February). In Korean.

LCH.Clearnet. Website www.lchclearnet.com.

McKibbin, Warwick J., and Peter J. Wilcoxen. 2002. The Role of Economics in Climate Change Policy. *Journal of Economic Perspectives* 16 (2): 107–29.

Montgomery, W. David. 1972. Markets in Licenses and Efficient Pollution Control Programs. *Journal of Economic Theory* 5 (3): 395–418.

Murray, Brian C., Richard G. Newell, and William A. Pizer. 2009. Balancing Cost and Emissions Certainty: An Allowance Reserve for Cap-and-Trade. *Review of Environmental Economics and Policy* 3 (1): 84–103.

Newell, Richard G., and Robert N. Stavins. 2003. Cost Heterogeneity and the Potential Savings from Market-Based Policies. *Journal of Regulatory Economics* 23 (1): 43–59.

orbeo. Website www.orbeo.com.

Pizer, William A. 2002. Combining Price and Quantity Controls to Mitigate Global Climate Change. *Journal of Public Economics* 85 (3): 409–34.

Presidential Committee on Green Growth. 2009a. The Law for Low Carbon, Green Growth; a bill submitted to Congress, February 2009. In Korean.

Presidential Committee on Green Growth. 2009b. *Green Growth for the Next Five Years*. Seoul: Presidential Committee on Green Growth. In Korean.

Presidential Committee on Green Growth. 2009c. *Three Mitigation Scenarios for the 2020 Midterm National GHG Mitigation Target*. Seoul: Presidential Committee on Green Growth. In Korean.

Prime Minister's Office, Climate Change Office. 2008. *Comprehensive Plan on Combating Climate Change*. Seoul: Climate Change Office of the Prime Minister's Office. In Korean.

Sorrell, Steven, and Jos Sijm. 2003. Carbon Trading in the Policy Mix. *Oxford Review of Economic Policy* 19 (3): 420–37.

Stavins, Robert N. 1995. Transaction Costs and Tradeable Permits. *Journal of Environmental Economics and Management* 29 (2): 133–48.

Stavins, Robert N. 2009. The Wonderful Politics of Cap-and-Trade: A Closer Look at Waxman-Markey. Article posted 27 May 2009 on *An Economic View of the Environment*, Belfer Center for Science and International Affairs, Harvard University. Available at belfercenter.ksg.harvard.edu.

Stern, Nicholas. 2007. *The Economics of Climate Change: The Stern Review*. Cambridge and New York: Cambridge University Press.

Weitzman, Martin. 1974. Price vs. Quantities. *Review of Economic Studies* 41 (4): 477–91.

Index